I0466151

Electric Telegraph Series

THE ELECTRIC TELEGRAPH

POPULARISED

NUMBER 4 IN THE ELECTRIC TELEGRAPH SERIES

DIONYSIUS LARDNER

THE ELECTRIC TELEGRAPH
POPULARISED

TGR RENASCENT BOOKS
2019

PUBLICATION HISTORY

THE ELECTRIC TELEGRAPH
POPULARISED

DIONYSIUS LARDNER
1855

Third Edition

REVISED AND RE-WRITTEN
by
Edward B. Bright
1867

This new edited edition published by
TGR Renascent Books
27 Springdale Court
Mickleover, Derby DE3 9SW
United Kingdom
2019

www.renascentbooks.co.uk

ISBN 978-1-7963481-4-9

CONTENTS

1. Subjugation of the powers of nature to human uses—2. Locomotion twenty years since—3. Circulation of intelligence then—4. Supposed prediction of succeeding improvements—Railway locomotion—5. Electric telegraphy—6. Fabrication of diamonds—sun-pictures—gas-lighting—electro-metallurgy—7. Such predictions would have been deemed incredible—8. Electro-telegraphy the most incredible of all—9. Remarkable experiment by Messrs. Leverrier and Lardner—10. Velocity of electric current—11. No limit to the celerity of telegraphy—12. Physical character of electricity—13. Not essential to the explanation of electro-telegraphy—14. Electricity a subtle fluid—15. Properties available for telegraphy—16. Voltaic battery—17. It is to the electric telegraph what the boiler is to the steam-engine—18. Means of transmitting the fluid in required directions—19. Conductors and insulators—20. Conducting wires—21. Voltaic battery—22. Transmission and suspension of the current—23. Current established by earth contact—24. Theories of earth contact—25. The return of the current through the earth—26. Various bodies evolve electricity—27. Common plate battery of zinc and copper—28. Why zinc and copper are preferred—29. Charcoal substituted for copper—30. Elements not essential—31. Strength of current varied by number of cells—32. Velocity of the electric current under different conditions—33. Production of telegraph signals

34. Relation of the deflection to the direction of the current—35. Galvanometer or multiplier—36. Method of covering the wire—37. Method of mounting the needle—38. Method of transmitting signals by the galvanometer—39. How the current may produce a temporary magnet—

Contents

40. Electro-magnets acquire and lose their magnetism instantaneously—41. How magnetic pulsations are rendered visible and counted—42. Extraordinary celerity of the oscillations thus produced—43. They produce musical sounds by which the rate of vibration may be estimated.—44. How the vibrations may impart motion to clockwork—45. Their action on an escapement—46. How the movement of one clock may be transmitted by the current to another—47. How an electro-magnet may produce written characters on paper at a distant station—48. How the motion of the hand upon a dial at one station can produce a like motion of a hand upon a dial at a distant station—49. How an agent at one station can ring an alarum at another station—50. Or may discharge a gun or cannon there—51. Power of the bell or other signal not dependent on the force of the current—52. Mechanism of telegraphic alarum—53. Various alarums in telegraphic offices—54. Magneto-electricity—55. Method of producing a momentary magneto-electric current—56. Application of an electro-magnet to produce it—57. Momentary currents alternately in contrary directions—58. Method of producing momentary currents all in the same direction—59. Magneto electric machine—60. Chemical property of the electric current.

61. Early Telegraphing by preconcerted signals—62. Semaphore telegraphs—63. Experiments with frictional or machine electricity—64. Telegraphs worked by frictional electricity — This class of electricity too volatile—65. Applications of voltaic or chemical electricity in early telegraphs. — Soemmering, Schilling, &c.—66. Gauss and Weber's telegraph—67. W. F. Cooke's first telegraph—68. Cooke and Wheatstone's five-needle telegraph.

69. Common plate battery—70. Combination of currents—71. Loss of power by imperfect conduction—72. Cylindrical batteries—73. Pairs of battery plates, or cells, and poles defined—74. Origin of term voltaic pile—75. Use of sand in charging batteries—76. To vary intensity of current 77. Description of batteries—Wollaston's, Hare's, Daniel's, Grove's, Smee's, Bunsen's, Marie Davy's.

78. Zinc coating applied to iron wire to prevent rust—79. Injurious effects of smoke in towns—80. Reasons for insulation—81. Best material

Contents

Contents

Contents

LIST OF PLATES AND FIGURES

List of Plates and Figures

List of Plates and Figures

PUBLISHER'S NOTE

The international telegram service in Britain, inaugurated by private enterprise in 1845 but soon taken over by the General Post Office (GPO), and latterly by British Telecom (BT), ended in 2003. In the United States the service finished when Western Union sent its last telegram in 2006. As a consequence, most people today, in the age of the Internet, Satellite Communications, Mobile Phones, E-mail, Instant Text Messaging and Fibre-Optic Cable, have no idea that the world was once girdled with thin iron wires strung on poles over thousands of miles of often inhospitable terrain, or that armoured cables lay fathoms deep in abysmal darkness on the bottom of the oceans. It was via these fragile threads that the world once communicated.

The purpose of this Electric Telegraph Series is to publish in new editions some of the many books on telegraphy that first appeared in the Victorian era. Neglected and forgotten, dismissed as no longer relevant, these books are a treasure trove for historians of technology, research students and interested lay persons. The technology and operation of the telegraph very quickly achieved a level of development and sophistication that now seems quite staggering, as a perusal of the books in this series will soon show.

One caveat must be mentioned—the men writing these books were in complete ignorance of the *nature* of electricity, although of course fully conversant with its *effects*. Electricity was often called a "mysterious fluid," by Victorians on the analogy that electricity somehow flowed through a wire like water flows through a pipe. It was not until 1897 that the atom was "split" and J. J. Thomson discovered the electron, the sub-atomic particle that is ultimately responsible for the flow of electricity. It was well into the twentieth century before a coherent theory of electricity was developed and promulgated. Be cautious, therefore, when reading early authors on electricity. Readers who have the need should consult professionals or up-to-date text-books on the subject. For everyone else, the books in this series will provide wonderfully readable and easy to understand accounts of electricity, which while not always strictly accurate according to modern understanding, nevertheless supply everything needed to understand telegraphy, telegraphic circuits, telegraphic instruments and their ubiquitous power sources—hand-turned generators or wet batteries of exceptional size.

SERIES EDITOR'S FORWARD

The author of this book, Dionysius Lardner, was an Irish scientific writer and public lecturer who popularised science and technology, and is especially remembered today as the editor of the 133 volume *Cabinet Cyclopædia* – an encyclopaedic collection of scientific writings by many of the ablest savants of the day, including Lardner himself who contributed articles on arithmetic, geometry, heat, hydrostatics and pneumatics, mechanics and the newest science of all – electricity.

Dionysius Lardner
by Alexander Craig

EARLY LIFE IN DUBLIN

Dionysius Lardner was born in Dublin on 3 April 1793, the son of William Lardner, a solicitor in Dublin, who wished his son to follow the same calling. After some years of uncongenial desk work, Lardner entered Trinity College, Dublin, in 1812, and obtained a B.A. in 1817 and an M.A. in 1819, winning many prizes. He married Cecilia Flood on 19 December 1815, but they separated in 1820 and were divorced in 1835. About the time of the separation, he began a relationship with a married woman, Anne Maria Darley Boursiquot, the wife of a Dublin wine merchant. It is believed that he fathered her son, Dion Boucicault, the actor and dramatist, not least because he provided him with financial support until 1840. Whilst in Dublin, Lardner began to write and lecture on scientific and mathematical matters, and to contribute articles for publication by the Irish Academy. Although in holy orders Lardner never held a clerical living. Until 1827 he lived in Dublin, in the early stages of a career as a scientific writer and lecturer, contributing articles to the *Transactions of the Irish Academy*. He published a variety of treatises, mostly on mathematical subjects, intended for beginners in arithmetic, geometry, and calculus. Lardner was appointed the first professor of natural philosophy and astronomy at the new London University in 1827, at which time he moved to London.

CAREER IN LONDON

In 1828 Lardner was elected professor of natural philosophy and astronomy at University College, London, a position he held until he resigned his professorship in 1831. Lardner showed himself to be a successful populariser of science and became a significant participant in the autodidactic and scientific cultures centred in the capital. At the height of his career, he lectured extensively on scientific and technical subjects, wrote prolifically, and became a scientific celebrity. He wrote treatises on mechanics, pneumatics, and Newton's optics for the Society for the Diffusion of Useful Knowledge, and contributed the chapters on algebra and geometrical analysis to the *Encyclopedia Metropolitana*. He oversaw seven editions of *The Steam Engine Familiarly Explained and Illustrated*, which eventually had eleven editions in Britain, as well as American editions and translations into French, German, Italian, and Danish; *The Athenaeum* called it "the most popular mechanical treatise ever published". He promoted the new

Dionysius Lardner
by Edith Fortunée Tita de Lisle

Monthly Chronicle and edited it in 1838 and 1839. He contributed important articles to the *Edinburgh Review*, notably a clear account of Charles Babbage's calculating engine (which Babbage himself recommended), and discussions of issues relating to steam transportation. He enjoyed great popularity as a lecturer, appearing at the Royal, London, and London mechanics' institutions; his lectures to provincial literary and philosophical societies and mechanics' institutes were eagerly awaited and widely praised as appearances of a great figure from the metropolis. Some of his lectures took the form of short courses on such general topics in the physical sciences as pneumatics or mechanics, but his forte was the single lecture on a topic of contemporary moment, delivered in a popular style to a general audience. One such lecture at the Royal Institution attracted the largest audience of the decade to the same hall where Faraday established his fame; on two occasions (1834 and 1835) he delivered popular evening lectures at British Association meetings, discussing Babbage's calculating machines and steam locomotion.

INVOLVEMENT IN SCANDAL

In 1840 Lardner's career received a major setback as a result of his involvement with Mary Spicer Heaviside, the wife of Captain Richard Heaviside, of the Dragoon Guards. Learning nothing about the tribulations of dalliance with married women after his experience in Dublin, Lardner ran off to Paris with Mrs Heaviside, pursued by her husband. When he caught up with them, Heaviside subjected Lardner to a beating, but could not induce Mary to return with him to England. Later that year he successfully sued Lardner for 'criminal conversation' (adultery) and received a judgment of £8,000. The Heavisides were divorced in 1845, and in 1846 Lardner was able to marry Mary Heaviside. They remained together for the rest of his life. The scandal caused by his affair with a married woman however, effectively ended his career in England.

FINAL YEARS IN PARIS

Lardner and his wife Mary remained in Paris until shortly

before his death, during which period they moved to Italy. After the 1840 scandal interrupted his London career he lectured widely in the United States from 1841 to 1844. Although he reputedly prospered from both lecture fees and the subsequent publication of his American lectures, earning from them around £40,000, the American sojourn brought him no scientific notice. **He died in Naples, Italy, on 25 April 1859, aged 66, and is buried there in the Cimitero degli Inglesi.**

LARDNER AND THE ELECTRIC TELEGRAPH

Unlike most, if not all, of the early writers who set out to explain the new science and technology of electrical communications to lay audiences, the author of this book never had any professional involvement with the telegraph. Readers of this book therefore, might well be astonished at the breadth and depth of knowledge about the subject that he reveals in this treatise. He discusses the theory and technology of the electric telegraph authoritatively, he describes in detail the uses to which it was put and the costs of doing so. He surveys the operation of telegraphs in Great Britain, the United States of America, Russia and in most of the European countries. He does not omit to explain, in a separate chapter, the somewhat specialised subject of railway telegraphy. We can only be amazed at the diligence and dedication required to amass this amount of material, to study and master its purport, and then present it in a highly readable form. As an important mediator of the culture of the new technologies of his time, Lardner's influence should not be underestimated, and his writings are a key source for understanding nineteenth-century popular ideas about progress and its relation to technological development. Although the day of the electric telegraph is now long past, supplanted by modern versions of the telegram and cablegram such as text messaging and e-mail, we can still enjoy and be enlightened by the writings of this remarkable man on a fascinating if obsolete technology.

Gordon Roberts
Derby 2019

THE

ELECTRIC TELEGRAPH

POPULARISED.

WITH ONE HUNDRED ILLUSTRATIONS.

By DIONYSIUS LARDNER, D.C.L.,

Formerly Professor of Natural Philosophy and Astronomy in University College, London.

FROM

"THE MUSEUM OF SCIENCE AND ART."

LONDON:

WALTON AND MABERLY,

UPPER GOWER STREET AND IVY LANE, PATERNOSTER ROW.

1855.

Facsimile of the first edition 1855 title page.

THE

ELECTRIC TELEGRAPH

BY

DR. LARDNER

𝔄 𝔑𝔢𝔴 𝔈𝔡𝔦𝔱𝔦𝔬𝔫

REVISED AND RE-WRITTEN

BY

EDWARD B. BRIGHT, F.R.A.S.

SECRETARY OF THE BRITISH AND IRISH MAGNETIC TELEGRAPH COMPANY

WITH 140 ILLUSTRATIONS

LONDON

JAMES WALTON

BOOKSELLER AND PUBLISHER TO UNIVERSITY COLLEGE

137, GOWER STREET.

1867.

Facsimile of the third edition 1867 title page.

AUTHOR'S PREFACE

Preface to the 1855 Edition

In the composition of this volume my purpose has been to render intelligible to all who can read, irrespective of any previous scientific acquirements, the various forms of telegraph in actual operation in different parts of the world, and the manner in which their marvellous effects are produced. Since the instrument in one form or another involves all the great laws governing electrical and magnetical phenomena, the discovery of which will render for ever memorable the researches of the eminent scientific men of the last half century, it was necessary to include in the exposition of each piece of apparatus such an account of the physical principle upon which its use depends, as should render its application and effects understood. Descriptions of such apparatus, however clearly expressed, would have been obscure without graphic illustrations to correspond with them. These have accordingly been supplied, as will be seen, with no sparing hand.

No two countries agree in adopting the same form of telegraphic instrument, and even in the same country different forms of telegraph are used by different companies and for different purposes. Since these various instruments are always different in the details of their construction and often totally distinct in their principle and mode of operation, it was necessary to explain each in succession, and to do so correctly it was necessary to seek and obtain authentic documents, descriptions, and drawings from those who were placed in the direction and superintendence of the telegraphs in various parts of the continent of Europe and in the United States.

The reader of this little volume will find in its pages abundant evidence that no pains or cost have been spared in these researches. The history of the invention of the Electric Telegraph is a subject upon which I have not judged it expedient to enter. The details of such a narrative, necessarily numerous and complicated, involving several questions of disputed priority and contested claims, besides filling a much larger volume than the present, would present few attractions for the large masses to whom our work is addressed.

The Electric Telegraph is not the invention of an individual. As it now exists, it is the joint production of many eminent scientific men and distinguished artists of various countries, whose labours and experimental researches on the subject have been spread over the last twenty years. Not being prepared to engage in a complete account of the progressive results of these labours, I have in the following work generally abstained from the mention of inventors, from a desire to avoid the risk of appearing to put forward some in undue preference to others who might be supposed to have better claims to notice. There can, however, be no risk of committing an injustice by stating that in England Professor Wheatstone, in the United States Professor Morse, in Bavaria M. Steinheil, in Prussia Dr. Siemens, and in France MM. Breguet and Froment, have severally stood in the leading ranks of invention. Besides these eminent persons may be mentioned, Mr. Bain, the inventor of the electro-chemical telegraph; Mr. Henley and the Messrs. Bright, who have improved the magnetic telegraph; Messrs. Brett, to whose genius and enterprise the world is indebted for submarine telegraphs; Messrs. Newall and Co., who have been signalised by the construction of submarine cables; Mr. Walker of the South Eastern Telegraph Company; and Mr. House of the United States, the inventor of a printing telegraph in extensive operation.

D. L.

1855

PREFACE TO THE THIRD (1867) EDITION

When the writer assisted the late Dr. Lardner in preparing the first edition of this treatise, thirteen years ago, telegraphy was comparatively in its infancy. Many inventions which at that time appeared important are now obsolete; while a great variety of appliances have since been brought to bear that were not then dreamt of. The whole system of electric communication has, in fact, become a recognised science in its various branches, and the laws governing the development and application of this wonderful power have been carefully investigated.

It may now be said that wherever civilisation exists, there telegraphic communication has penetrated, linking up continents, countries, and races. The mission of the sentient wire is a noble one: to bind together nations, and colonies to their mother countries. Great Britain is now joined to India on the one hand, and to Canada, the United States, and even Vancouver's Island, on the other; our cousins at the antipodes in Australia will shortly be added; and that important stepping-stone to the East, Malta, has long been put within prompt hail of England as well as Egypt; — Gibraltar is to follow.

The British dominions thus become united in instant correspondence, notwithstanding the barriers interposed by half our planet's surface, and the vast oceans intervening. In no applied science has the advance been so rapid as in the Electric Telegraph. This means of conveying thought in a moment between distant points, which twenty years ago had scarcely emerged in a practical form from the philosopher's studio, now covers the world with its vast network of communicating fibres. In attaining such a result, England has stood in the foreground. To her is due not only the origination of the submarine telegraph, but she has also provided the capital, constructed the cables, and carried out the laying of every important submarine communication.

On land the telegraph has had many difficulties to encounter. In countries like England and Switzerland the humidity of the climate interferes with insulation; in India monkeys, ants, and heat have proved causes of mischance; in the deserts the Arabs, and on the prairies Indians have had to be bribed or coerced. In Russia, again, snows, and in the forests of America falling trees, have interrupted for a time the working of the apparatus. Hence various countries have their different systems of telegraph, each more or less designed to meet the special conditions attending their erection and working.

The descriptions contained in the former editions could not therefore convey any adequate idea of present progress or telegraphic apparatus, and it has become necessary to re-model and re-write the work. As far as possible, only practical points are dealt with, and the object has been to explain them in a popular and at the same time a condensed form.

Among the new branches of the subject now introduced, may be mentioned the Atlantic Telegraph, the line to India, the Malta and Alexandria, and other important submarine works; the greatly improved contrivances for train signalling on railways; the regulation of public clocks by electricity, now extensively adopted in London, Liverpool, Glasgow, Edinburgh, &c.; and the system of meteorological signals and storm warnings introduced by the late Admiral Fitzroy, and carried out by the Board of Trade.

E. B. B.

January, 1867.

PLATE I

Double needle telegraph instrument
(the enclosure on the top hides the alarum)

1

THE THEORY, PROPERTIES AND VELOCITY OF ELECTRICITY

1. EACH succeeding age and generation leaves behind it a peculiar character, which stands out in relief upon its annals, and is associated with it forever in the memory of posterity. One is signalised for the invention of gunpowder, another for that of printing; one is rendered memorable by the revival of letters, another by the reformation of religion; one is marked in history by the conquests of Napoleon, another is rendered illustrious by the discoveries of Newton.

If we are asked by what characteristic the present age will be marked in future records, we answer, by the miracles which have been wrought in the subjugation of the powers of the material world to the uses of the human race. In this respect no former epoch can approach to competition with it.

The author of some of the most popular fictions of the day has affirmed, that in adapting to his purpose the results of his personal observation on men and manners, he has not infrequently found himself compelled to mitigate the real in order to bring it within the limits of the probable. No observer of the progress of the arts of life, at the present time, can fail to be struck with the prevalence of the same character in their results as that which compelled this writer to suppress the most wonderful of what had fallen under his eye, in order to bring his descriptions within the bounds of credibility.

2. Many are old enough to remember the time when persons, correspondence, and merchandise were transported from place to place in this country by stage-coaches, vans, and wagons. In those days the fast-coach, with its team of spanking blood-horses, and its bluff driver, with broad-brimmed hat and drab box-coat, from which a dozen capes were pendant, who "handled the ribbons" with such consummate art, could pick a fly from the ear of the off-leader, and turn into the gateway at Charing Cross with the precision of a geometrician, were the topics of the unbounded admiration of the traveller. Certain coaches obtained a special celebrity and favour with the public. We cannot forget how the eye of the traveller glistened when he mentioned the Brighton "Age," the Glasgow "Mail," the Shrewsbury "Wonder," or the Exeter "Defiance," — the "Age," which made its trip in five hours, and the "Defiance," which acquired its fame by completing the journey between London and Exeter in less than thirty hours.

3. The rapid circulation of intelligence was also the boast of those times. How foreigners stared when told that the news of each afternoon formed a topic of conversation at tea-tables the same evening, twenty miles from London, and that the morning journals, still damp from the press, were served at breakfast within a radius of thirty miles, as early as the frequenters of the London clubs received them.

Now let us imagine that some profound thinker, deeply versed in the resources of Science at that epoch, were to have gravely predicted that the generation existing then and there would live to see all these admirable performances become obsolete, and consigned to the history of the past; that they would live to regard such vehicles as the "Age" and "Defiance" as clumsy expedients, and their celerity such as to satisfy those alone who were in a backward state of civilisation!

4. Let us imagine that such a person were to affirm that his contemporaries would live to see a coach like the "Defiance" making its trip between London and Exeter, not in thirty, but five hours, and drawn, not by 200 blood-horses, but by a moderate sized stove and four bushels of coals!

5. Let us further imagine the same sagacious individual to predict that his contemporaries would live to see a building erected in the centre of London, in the cellars of which machinery would be provided for the fabrication of *artificial lightning*, which should be supplied *to order*, at a *fixed price*, in any quantity required, and of *any prescribed force*; that *conductors* would be carried from this building to all parts of the country, along which such *lightning* shall be sent at will; that in the attics of this same building would be provided certain small instruments like barrel-organs or pianofortes; that by means of these instruments, the aforesaid lightning should, at the will and pleasure of those in charge of them, deliver messages at any part of Europe, Asia, or America, from St. Petersburg in the north, to Calcutta or Rangoon in the east, to New York, San Francisco, or even Vancouver's Island in the west; and, in fine, that answers to such messages should be received instantaneously, and by like means: that in this same building, offices should be provided, where any lady or gentleman might enter, at any hour, and for a few shillings send a message by *lightning* to Paris or Vienna, and by waiting for a few moments, receive an answer!

Might he not exclaim, after the inspired author of the book of Job:—

"Canst thou send lightnings, that they may go, and say unto thee, Here we are?" xxxviii., 35.

But, suppose that his foresight should further enable him to pronounce that means would be invented by which any individual in any one town or city of Europe should be enabled to take in his hand a pencil or pen, the point of which should be in any other town or city, no matter how distant, and should, with such pen or pencil, write or delineate in such distant place, such characters or designs as might please him, with as much promptitude and precision as if the paper to which these characters or designs were committed lay upon the table before him; or that an individual pulling a string at London should ring a bell at Vienna, or holding a match at St. Petersburg should discharge a cannon at Naples!

6. Suppose he should affirm that means would be discovered for converting charcoal into diamonds; that the light of the sun would be compelled, without the intervention of the human hand, to make a portrait or a picture, with a fidelity, truth and precision, with which the productions of the most exalted artistic skill would not bear comparison; and that this picture should be produced and completed in its most minute details in a few seconds — nay, even in the fraction of a second; that candles and lamps would be superseded by flame manufactured on a large scale in the suburbs of cities, and distributed for use in pipes, carried under the streets, and into the houses and other buildings to be illuminated; and that the precious and other metals being dissolved in liquids, would form themselves into the articles of ornament and use by a spontaneous process, and without the intervention of human labour!

No authority however exalted, no attainments however profound, no reputation however respected, could have saved the individual rash enough to have given utterance to such predictions some forty years ago, from being regarded as labouring under intellectual derangement. Yet all these things have not only come to pass, but the contemplation of many of them has become so interwoven with our habits, that familiarity has blunted the edge of wonder.

7. Compared with all such realities, the illusions of oriental romance grow pale; fact stands higher than fiction in the scale of the marvellous; the feats of Aladdin are tame and dull; and the slaves of the lamp yield precedence to the spirits which preside over the battery and the boiler.

8. Of all the physical agents discovered by modern scientific

research, the most fertile in its subserviency to the arts of life is incontestably electricity, and of all the applications of this subtle agent, that which is transcendently the most admirable in its effects, the most astonishing in its results, and the most important in its influence upon the social relations of mankind, and upon the spread of civilisation and the diffusion of knowledge, is the Electric Telegraph. No force of habit, however long continued, no degree of familiarity can efface the sense of wonder which the effects of this most marvellous application of science excites.

9. Being at Paris in 1850, Dr. Lardner was engaged to share with M. Leverrier, the celebrated astronomer, and some other men of science, in the superintendence of a series of experiments to be made before committees of the Legislative Assembly and of the Institute, with the view of testing the efficiency of certain telegraphic apparatus.

Two wires, extending from Paris to Lille, were united at the latter place, so as to form one continuous wire, extending to Lille and back, making a total distance of 336 miles. This, however, not being deemed sufficient for the purpose, several coils of wire wrapped with silk were obtained, measuring in their total length 746 miles, and were joined to the extremity of the wire returning from Lille, thus making one continuous wire measuring 1082 miles. A message consisting of 282 words was then transmitted from one end of the wire. A pen attached to the other end immediately began to write the message on a sheet of paper moved under it by a simple mechanism, and the entire message was written in full in the presence of the Committee.

This might well be looked upon as a feat sixteen years ago, but the science of Telegraphy has made such wonderful progress, that at the present time the two Atlantic cables, when joined end to end, so as to form one unbroken length of nearly 4000 miles, can readily be worked at a speed far greater than through the comparatively small length of line referred to in the foregoing experiment. Each cable can easily pass fifteen words per minute, or upwards of forty average messages per hour between Europe and America! This speed of transmission is greatly exceeded in the working of short land lines.

But it must not be imagined, because we have here produced an example of the transmission of despatches to a distance of 4000 miles, that any augmentation of that distance could cause any delay of practical importance, provided the electric current was not allowed to ooze out, and so diminish in power, during its transmission.

10. Although the velocity of the electric current has not been very exactly measured, it has been established beyond all doubt that it is so great that to pass from any one point on the surface of the earth to any other, it would take no more than a fraction of a second.

11. If, therefore, the despatch had been sent to a distance of 20,000 miles, its transmission would still have been instantaneous.

Such a despatch would fly round the earth between the two beats of a common clock, and would be written in full at the place of its destination more rapidly than it could be repeated by word of mouth. When such statements are made, do we not feel disposed to exclaim:–

> "Are such things here as we do speak about?
> Or have we eaten of the insane root,
> That makes the reason prisoner?"

In its wildest flights the most exalted imagination would not have dared, even in fiction, to give utterance to these stubborn realities. Shakespeare only ventured to make his fairy

> "Put a girdle round the earth
> In forty minutes."

To have encircled it in a second, would have seemed too monstrous, even for Robin Goodfellow.

The curious and intelligent reader of these pages will scarcely be content, after the statement of facts so extraordinary, to remain lost in vacant astonishment at the power of science, without seeking to be informed of the manner in which the phenomena of nature have been thus wonderfully subdued to the uses of man. A very brief exposition will be enough to render intelligible the manner in which these miracles of science are wrought.

12. The World of Science is not agreed as to the physical character of Electricity. According to the opinion of some it is a fluid infinitely lighter and more subtle than the most attenuated and impalpable gas, capable of moving through space with a velocity commensurate with its subtleness and levity. Some regard this fluid as simple. Others contend that it is compound, consisting of two simple fluids having antagonistic properties, which, when in combination, neutralise each other, but which recover their activity by decomposition. Others again regard it not as a specific fluid which moves through space, but as a phenomenon analogous to sound, and think that it is only a series of undulations or vibrations that are propagated through a highly elastic medium which produce the various electrical effects, just as the pulsations of the atmosphere produce all the effects of sound.

13. Happily these difficult discussions are not necessary to the clear comprehension of the laws which govern the phenomena upon which electric telegraphy depends. It will nevertheless, for the purposes of explanation, be convenient to use a system of language which implies the existence of a certain fluid which we shall call the electric fluid, capable of moving through certain bodies, and being obstructed or altogether stopped by others, and which by its presence or proximity produces certain definite effects, mechanical and chemical.

14. Whether the electric agency be or be not a material fluid for our present purpose is unimportant. It is enough that it comports itself as such, and that the properties or effects which we shall impute to it are such only as it is ascertained by observation and experiment to possess or produce.

15. However various the forms may be which invention has conferred upon electric telegraphs, their efficiency in all cases depends on our power to produce at will the following effects:—

1st. To produce or develop the electric fluid in any desired quantity, and of the necessary quality.

2nd. To transmit it with celerity to any required distance, without injuriously dissipating it.

3rd. To cause it upon its arrival at any assigned point to produce some sensible effects which may serve the purpose of written or printed characters.

16. The electric fluid is deposited in a latent state in unlimited quantity in the earth, the waters, the atmosphere, and in all bodies upon the earth, whether solid, liquid, or gaseous. It is disengaged and rendered active by various causes, natural and artificial. The mutual friction of bodies, contact and pressure, the contiguity or contact of bodies having different temperatures, the chemical action of bodies one upon another, the action of magnetic bodies on each other, and on bodies susceptible of magnetism, are severally causes of the development of the electric fluid in greater or less quantity.

Founded upon these phenomena, various apparatus have been contrived, by means of which the electric fluid may be evolved and collected in any desired quantity, and with any required intensity. Among these, that which has proved to be the most efficient for telegraphic purposes is the GALVANIC or VOLTAIC BATTERY.

17. This apparatus is to the electric telegraph what the boiler is to the steam-engine. It is the generator of the fluid by which the action of the telegraphic machine is produced and maintained. It supplies the fluid in any desired quantity and of any desired intensity. As the boiler is supplied with expedients by which, within practical limits, the quantity and pressure of the steam may be varied according to the exigencies of the work to which the engine is applied, so the voltaic battery is provided with expedients by which the quantity and intensity of the electric fluid it evolves can be varied according to the distance to which the intelligence is to be transmitted; and the form, whether visible, oral, written or printed, in which it is to be required to be delivered at the place of its destination.

18. The electric fluid being thus produced in sufficient quantity, it is necessary to provide adequate means of transmitting it to a distance without exposing it to any cause of injurious dissipation or waste.

If tubes or pipes could be constructed with sufficient facility and cheapness, through which the subtle fluid could flow, and which would be capable of confining it during its transit, this object would be attained. As the galvanic battery is analogous to the boiler, such tubes would be analogous in their form and functions to the steam-pipe of a steam-engine.

19. The construction of such means of transmission has been accomplished by means of the well-known property of the electric fluid, in virtue of which it is capable of passing freely along a certain class of bodies called CONDUCTORS, while its movement is arrested by another class of bodies called NON-CONDUCTORS or INSULATORS.

The most conspicuous examples of the former class are the metals; the most remarkable of the latter being resins, wax, glass, porcelain, silk, cotton, dry air, &c.

20. Now if a rod or wire of metal be coated with wax, or wrapped with silk, the electric fluid will pass freely along the metal, in virtue of its character of a conductor; and its escape from the metal laterally will be prevented by the coating, in virtue of its character of an insulator.

The insulator in such cases is, so far as relates to the electricity, a real tube, inasmuch as the electric fluid passes through the metal included by the coating, in exactly the same manner as water or gas passes through the pipes which conduct it; with this difference, however, that the electric fluid moves along the wire more freely, in an almost infinite proportion, than does either water or gas in the tubes which conduct them.

If, then, a wire, coated with a non-conducting substance, capable of resisting the vicissitudes of weather, were extended between any two distant points, one end of it being attached to one of the extremities of a galvanic battery, a stream of electricity would pass along the wire — *provided the other end of the wire were connected by a conductor with the other extremity of the battery.*

21. How the fluid transmitted to a distant station is made to produce the effects by which messages are expressed will be explained hereafter; meanwhile it will be necessary first to explain the form and principle of the voltaic batteries used for telegraphic operations, and, secondly, the expedients by which the current is transmitted and suspended, and turned in one or another direction at the will of the operator at the station from which despatches are transmitted.

To comprehend the principle of the voltaic battery, let us suppose that two strips, cut, one *Z Z* from a sheet of zinc, and the other **C C** (Figure 1) from a sheet of copper, are immersed without touching each other in a vessel containing water slightly acidulated. To the upper edges **P** and **N** of the strips let two pieces of wire **P p** and **N n**, be soldered. In this state of the apparatus no development of the electric fluid will be manifested; but if the ends **p** and **n** of the wires be brought into contact, an electric current will set in, running on the wires from **P**, the point where the wire is soldered to the copper **C C**, to **N**, the point where the other wire is soldered to the zinc **Z Z**. This current will continue to flow so long as the ends **p** and **n** of the wires are kept in mutual

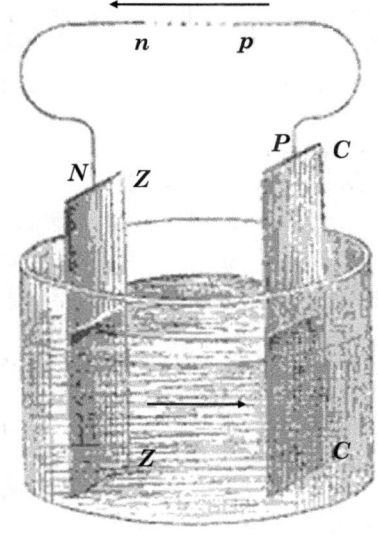

Fig. 1
A voltaic battery

contact, and no longer. The moment the ends **p** and **n** are separated, the current ceases.

22. The commencement of the current upon the contact of the wires, and its cessation upon their separation, are absolutely instantaneous; so much so, that if the ends **p** and **n** were brought into contact and separated a hundred times in a second, the flow and suspension of the current being simultaneous with the contacts and separations, would also take place a hundred times in a second.

The existence of the current established in this case is independent of the length of the wires **P p** and **N n**. Whether their length be 10 feet, 10 miles, or 100 miles, the current will still flow upon them when their extremities **p** and **n** are brought into contact. The only difference will be, that the intensity of the current will be less in the same proportion as the length of the wires is augmented.

23. There is another condition of great importance, whether regarded theoretically or practically, on which the current will be established and maintained.

Instead of bringing the wires **P p** and **N n** into contact, let them be continued downwards, as represented in Figure 2, and connected with two plates of metal **p'** and **n'**, buried in the earth, or with masses of metal or other good conducting body of any form whatever thus buried. Under these circumstances a new condition arises; the body of the earth being a general reservoir of electricity, receives and diffuses the surplus electricity from the positive or *plus* pole, and yields electricity to the negative or *minus* pole of the battery. This is analogous to the action of a pump drawing water from a well on the one hand and discharging it into a reservoir on the other.

24. In some of the electric theories, it was assumed that the course of the current must in all cases be continuous and unbroken from **P** to **N**, as it is in fact under the conditions represented in Figure 1, when the ends **p** and **n** are in contact. But it is generally now recognised that the process is one of equalisation when the battery poles or conducting wires are connected to the earth.

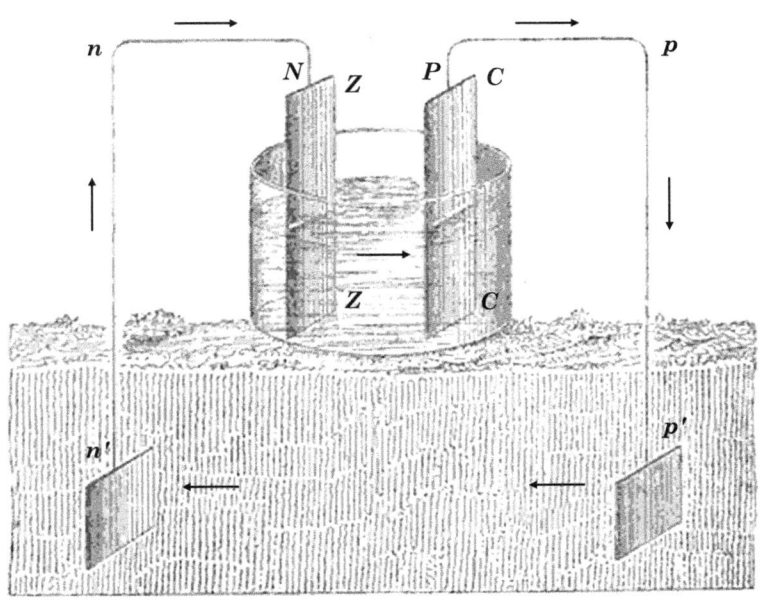

Fig. 2
Voltaic battery with earth connection

It is found also, in this case, that the existence of the current is independent of the lengths of the wires, which do not affect it otherwise than by diminishing its intensity. Whether the wires are 10 feet, 10 miles, or 100 miles in length, the currents still flow from *P* to *p'* and from *n'* to *N*.

25. Of all the miracles of science, surely this is the most marvellous. A stream of electric fluid has its source in the cellars of the Central Telegraph Office, London. It flows under the streets of the great metropolis, and, passing on wires suspended over a zigzag series of railways, reaches Edinburgh, where it dips into the earth, and diffuses itself from the buried plate.

Instead of burying plates of metal, it would be sufficient to connect the wires at each end with the gas or water pipes, which, being conductors, would equally convey the fluid to the earth; and in this case, every telegraphic despatch which flies to Edinburgh along the wires which border the railways would rush to the gas-pipes which illuminate Edinburgh, while the current from the opposite pole of the Lothbury batteries would be placed in connection with the gas-pipes which illuminate London.

26. To derive all the necessary instruction from what has been explained above, it will be necessary to distinguish what is essential from what is merely optional, and which admits of modification or change without affecting the result.

27. It will be seen that the electric fluid is evolved by the combination of three bodies, the zinc, the copper, and the acidulated solution in which they are immersed. The production of the current depends on the chemical action of the solution on the zinc. That metal being very susceptible of oxidation, decomposes the water which is in contact with it. One constituent of the water combining with the zinc, produces a compound called the oxide of zinc, and this oxide entering again into combination with the acid which the water holds in solution, forms a soluble salt. If the acid, for example, be sulphuric acid, this salt will be the sulphate of the oxide of zinc, and as fast as it is produced it will be dissolved in the water in which the slips of metal are immersed.

Meanwhile, the copper, not being as susceptible of chemical action as the zinc, remains comparatively unaffected by the solution;

but the hydrogen evolved in the decomposition of the water collects upon its surface, after which it rises and escapes in bubbles at the surface of the solution.

It is to this chemical action upon the zinc that the production of the electric current is due. If a like action had taken place in the same degree on the copper, a similar and equally intense electric current would be produced in the opposite direction; and in that case the two currents would neutralise each other, and no electric effect would ensue.

From this it will be seen that the efficacy of the combination must be ascribed to the fact, that one of the two metals immersed in the solution is more oxydizable than the other, and that the energy of the effect and the intensity of the current will be so much the greater as the susceptibility of oxidation of the one metal exceeds that of the other.

28. It appears, therefore, that the principle may be generalised, and that electricity will be developed, and a current produced by any two metals similarly placed, which are oxydizable in different degrees. And, indeed, if two pieces of the same metal are differently acted upon, either by heat or chemically, a current of electricity will be produced on their being connected together.

Zinc being one of the most oxydizable metals, and being also sufficiently cheap and abundant, is generally used by preference for voltaic combinations. Silver, gold, and platinum are severally less susceptible of oxidation, and of chemical action generally, than copper, and would therefore answer voltaic purposes better, but are excluded by their greater cost, and by the fact that copper is found sufficient for all practical purposes.

29. It is not, however, absolutely necessary that the unoxydizable element *C C* of the combination should be a metal at all. It is only necessary that it be a good conductor of electricity. In certain voltaic combinations, charcoal properly solidified has therefore been substituted for copper, the solution being such as would produce a strong chemical action on copper.

30. In the above illustration, we have supposed that the metallic elements of the combination are thin rectangular slips cut from the sheet metal. The form, however, is in no manner essential to the production of the electric current. So long as the magnitude of the surfaces exposed to contact with the solution is the same, the current will have the same force. The pieces of metal may therefore have the form here supposed of thin rectangular plates, or they may be formed, as is often found convenient, into hollow cylinders, that of the copper being so much less in diameter than that of the zinc, that it is capable of being placed within it without mutual contact.

The simple arrangement first adopted by Volta consisted of two equal discs of metal, one of zinc, and the other of copper or silver, with a disc of cloth or bibulous card, soaked in an acid or saline solution, between them. These were usually laid, with their surfaces horizontal, one upon the other.

31. To produce the effects, whatever these may be, by which the telegraphic messages are expressed, it is necessary that the electric current shall have a certain intensity. Now, the intensity of the current transmitted by a given voltaic battery along a given line of wire will decrease, other things being the same, in the same proportion as the length of the wire increases. Thus, if the wire be continued for 10 miles, the current will have twice the intensity which it would have if the wire had been extended to a distance of 20 miles.

It is evident, therefore, that the wire may be continued to such a length that the current will no longer have sufficient intensity to produce at the station to which the despatch is transmitted those effects by which the language of the despatch is signified.

The intensity of the current transmitted by a given voltaic battery upon a wire of given length, will be increased in the same proportion as the area of the section of the wire is augmented. Thus, if the diameter of the wire be doubled, the area of its section being increased in a fourfold proportion, the intensity of the current transmitted along the wire will be increased in the same ratio.

In fine, the intensity of the current may also be augmented by increasing the number of pairs of generating plates or cylinders composing the galvanic battery.

Since it has been found most convenient generally to use iron as the material for the conducting wires, it is of no practical importance to take into account the influence which the quality of the metal may produce upon the intensity of the current. It may be useful nevertheless to state that, other things being the same, the intensity of the current will be in the proportion of the conducting power of the metal of which the wire is formed, and that copper is the best conductor of the metals.

M. Pouillet found, by well-conducted experiments, that the current supplied by a voltaic battery of ten pairs of plates, transmitted upon a copper wire, having a diameter of four-thousandths of an inch, and a length of six-tenths of a mile, was sufficiently intense for all the common telegraphic purposes. Now, if we suppose that the wire, instead of being four-thousandths of an inch in diameter, has a diameter of a quarter of an inch, its diameter being greater in the ratio of 62½ to 1, its section will be greater in the ratio of nearly 4000 to 1, and it will consequently carry a current of equal intensity over a length of wire 4000 times greater, that is, over 2,400 miles of wire.

But in practice it is needless to push the powers of transmission to any such extreme limits. To reinforce and maintain the intensity of the current, it is only necessary to establish at convenient intervals along the line of wires intermediate batteries, by which fresh supplies of the electric fluid shall be produced; and this may in all cases be easily accomplished, the intermediate telegraphic stations being at distances, one from another, much less than the limit which would injuriously impair the intensity of the current.

It is, in fact, economical to distribute the battery power in this manner upon a line of telegraph, as the current set in motion on each section is much less intense, and therefore has not so great a tendency to escape at the points of insulation.

Having thus explained the means by which an electric current can be conducted from any one place upon the earth's surface to any other, no matter what be the distance between them, and how all the necessary or desired intensity may be imparted to it, we shall now proceed to explain the expedients by which such a current may enable a person at one place to convey instantaneously to another place, no matter how distant, signs serving the purpose of written language.

It may be shortly stated that the production of such signs depends on the power of the agent transmitting the current to transmit, suspend, intermit, divert, and reverse it at pleasure.

32. Although there is some discordance in the results of experiments made to determine the velocity of the current, they all agree in proving it to be prodigious. It varies according to the conducting power of the metal of which the wire is composed. On copper wire, its velocity, according to Professor Wheatstone's experiments, is 288,000 miles; and according to those of MM. Fizeau and Gonelle, 112,680 miles, per second. It should be borne in mind that these experiments were made with frictional (or machine) electricity, upon short lengths of wire in a room, and that this species of electricity has been found too volatile and evanescent to be employed practically for telegraphic purposes. On the iron wire used for telegraphic purposes, its velocity is 62,000 miles per second, according to Fizeau and Fonelle; 28,500 according to Professor Mitchell, of Cincinnati; and about 16,000 according to Professor Walker, of the United States. When, however, wires covered with gutta percha were brought into use and laid in considerable lengths underground, it was found that the transmission of the current through the wire was greatly reduced in speed. A series of experiments on this point were reported by the writer in a lecture at the British Association of 1854, showing the rate of passage of the currents through underground copper wires of No. 16 gauge thus insulated, to be about 1000 miles per second, the velocity varying somewhat with the character of the electric current employed. He referred this reduction of velocity to the analogy between a wire so circumstanced and the internal coating of a Leyden jar; the analogy resulting in the temporary retention of an electric charge by the gutta percha coated wire. It was also found that part of the electric charge communicated to the wire was, upon the withdrawal of the voltaic battery, quickly thrown back to the starting-point. This would follow in consequence of the electricity temporarily absorbed by the half-length of wire nearest the starting-point naturally finding it a shorter path of discharge to return whence it came.

These phenomena were most ably analysed by Dr. Faraday,

and completed the remaining link of comparison, then wanting, between what had been termed static and dynamic electricities.†

It is evident, therefore, that the interval which must elapse between the production of any change in the state of the current at one telegraphic station, and the production of the same change at any other on an overground line, cannot exceed a very minute portion of a second; and since the transmission of signals depends exclusively on the production of such changes, it follows that such transmission must be practically instantaneous.

In the transmission of signals through the 2,500 miles of gutta-percha-covered wire, forming the conductor of the Atlantic cable of 1858, intervals varying from 2¾ to 4 seconds were actually observed to elapse between the communication of the electric spark at the one end of the cable and its appearance at the other end. It was also, however, found that several currents of different kinds might be made to traverse the Atlantic wire in the same direction without interfering with the action of each other on arriving at the further end.

33. Since all telegraphic signals depend on the power of the agent who makes them, to transmit, control, and modify the current at will, it must be apparent how important it is for those who desire to understand this interesting subject, to comprehend, in the first instance, the means by which this power is obtained and exercised.

It is necessary to remember that the current will flow along a line of conducting wire so long as, and no longer than, a voltaic battery is interposed at some point on the line, the wire being attached to its poles, and the remote ends of the wire connected with the earth, as explained in **(23)**, and that, provided the battery have adequate force, it does not matter how distant from its poles the points may be at which the wires are connected with the earth.

If at any point of the line the wire is broken, the current instantly ceases along the entire line. If it be reunited, the current is instantly re-established. If the connection of the wire with the poles of the battery be reversed, so that the end which was connected with the positive is transferred to the negative pole, and *vice versa,*

---

† See "Report Brit. Assoc. 1854."

the direction of the current along the entire line is reversed — since it must always flow *from* the positive and *to* the negative pole. If at any point the wire, being broken, be connected with another wire proceeding to the earth in any other direction, the current will be diverted to the latter wire, deserting its former course. If the wire conducting the current be connected at the same point with two wires both connected with the earth, it will be distributed between the two, the greater part, however, following that wire which offers the easier road to the earth.

These few principles, which are clear and simple, supply an easy key to the whole art of electro-telegraphy.

The manner in which the pulsations of the current are produced, controlled, and regulated by the operator at a station being understood, it will next be necessary to show how they are made to produce signals at the station to which the despatch is transmitted, by which the operator or observer there can be enabled to understand and interpret the communication.

The effects of the current which have been found most con-venient for this purpose, are:–

1st. Its power to deflect a magnetic needle from its position of rest, and to throw it into another direction.

2nd. Its power to impart temporary magnetism to soft iron, this magnetism suddenly deserting the iron when the current is suspended.

3rd. Its power to produce the chemical decomposition of certain substances.

All forms of electric telegraph depend on one or other of these properties of the current.

If a wire be extended over and under a compass-needle which directs itself to the magnetic north and south, parallel to the needle, and as close to it as it can be placed without actually touching it, as represented in Figure 3, the needle will remain undisturbed

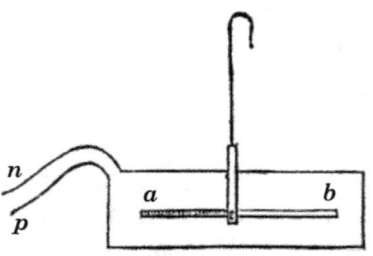

Fig. 3
Compass needle encircled by a wire carrying a current

in its position. Let the ends **p** and **n** of the wire be then attached to the poles of a voltaic battery, so that a current of a certain intensity shall be transmitted upon it. The moment the current is established upon the wire, the magnetic needle **a b** will be thrown out of its usual direction, and instead of pointing north and south, it will point east and west. If the direction of the current upon the wire be reversed, the direction of the deflection of the needle will be reversed.

PLATE II

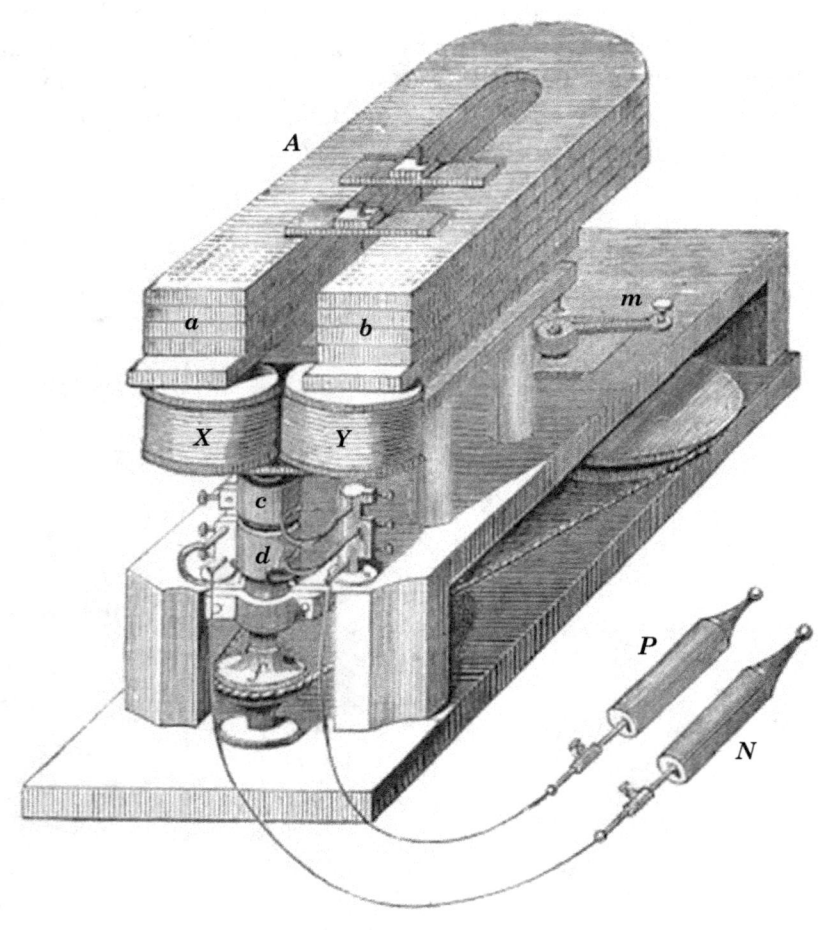

Magneto-electric machine
(a description of this apparatus occurs on p.47)

2

THE VARIOUS ARRANGEMENTS BY WHICH ELECTRICITY IS MADE TO PRODUCE MOTION AND MECHANICAL ACTION

34. TO explain the manner in which the deflection of the magnetic needle depends on the direction of the current, let us suppose the needle to be placed on an horizontal axis *O*, Figure 4, so as to play in a vertical plane, and to be maintained in the vertical direction when not affected by the current, by giving a slight preponderance to the arm on which the south pole of the needle is placed. By this arrangement the needle, when undisturbed, will rest in the vertical position, the north pole *N* being directed upwards, and the south pole *S* being directed downwards.

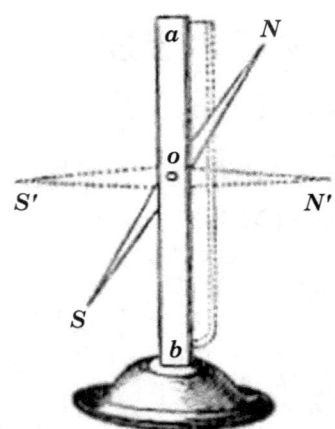

Fig. 4
Deflection of compass needle

Now if the current which is before the needle be directed *downwards* and that which is behind it *upwards*, the *north* pole **N** will be deflected to the *right*, and consequently the *south* pole **S** to the *left*, as represented in the figure. But if the direction of the current be reversed so that *before* the needle, it shall be directed *upwards* and *behind* it *downwards*, the north pole **N** will be deflected to the left and the south pole **S** to the right.

If the intensity of the current be great, and the preponderance given to the lower arm of the needle small, the deflective force of the current will be sufficient to throw the needle completely at right angles to its position of rest, that is, to give it the horizontal direction; but it is important to observe, that no greater intensity of the current can affect it further. The north pole, for example, cannot be deflected downwards, or the south pole upwards. In fine, the needle cannot be more affected by any increase of force of the current after it has once been thrown into the horizontal direction.

If the intensity of the current be insufficient to throw the needle into the horizontal direction, it will nevertheless take a position intermediate between that and the vertical direction at which it will rest. Its deflection from the vertical will be more and more considerable as the current is more intense, and certain mathematical conditions have been discovered by which the relative intensity of the current may be determined by the amount of the deflection of the needle which it produces.

35. It is evident that the sensibility of the needle will be so much the greater as the preponderance of the arm **S** is diminished and the intensity of the current increased. An expedient has, however, been ingeniously contrived, by which the most feeble current can be made to affect the needle. This is accomplished by winding the wire which carries the current several times round the needle, each coil being still parallel to the needle. By this contrivance, each successive coil of the wire produces a separate effect upon the needle, and if there be fifty such coils passing successively before and behind the needle, each portion of the wire thus carrying the current producing an independent deflecting force, there will be a total deflecting force a hundred times greater than that which a single portion of the wire passing once over or under the needle would produce.

In this manner the deflecting power of the most feeble current may be so *multiplied* as to produce upon the needle as powerful an effect as would be produced by a current of great intensity.

An apparatus consisting of wire thus coiled round a magnetic needle is called a MULTIPLIER, inasmuch as it multiplies the deflecting power of the needle. It is also called a REOSCOPE, or REOMETER,† and sometimes a GALVANOSCOPE, or GALVANOMETER, inasmuch as it indicates the presence, and by certain arrangements measures the intensity, of a galvanic or voltaic current.

36. When the conducting wire is thus coiled round a needle, it is necessary that it should be covered or coated by some substance which is a non-conductor of electricity, since otherwise the coils being necessarily in contact one with another, the current, instead of following the continuous thread of wire, would pass from coil to coil. In such cases, therefore, the wire is wrapped with silk or cotton, which being a non-conductor, confines the current within it just as water would be included in a pipe.

37. As the wire coiled in the manner above described, and the frame which carries it, would prevent the play of the needle from being easily and conveniently observed, the needle included within the frame is fixed upon the axis which supports it, so that the axis turns with it. This axis passes through the side of the frame, on which the wire is coiled, and upon the end of it which projects beyond the frame an index hand is fixed, so as to be parallel to the needle, the play of which will necessarily correspond with that of the needle. This hand plays upon a sort of dial, by which its deviations to the right or to the left from its position of rest are indicated.

† From two Greek words—reos, meaning a current, and metron, a measure.

This will be more clearly understood by reference to Figure 5, which represents a section of the mounting of the needle, the coil of wire and their appendages, made by a vertical plane through the axis of the needle. The needle within the coil is represented at *a b*, in its position of rest. The axis of the needle passing through the frame supporting the coil, and through the dial plate, supports in front of the dial the hand *a' b'*, which is fixed upon the axis in a position parallel to the needle *a b*, so that it must play before the dial in a manner corresponding exactly with the play of the needle *a b* within the coil.

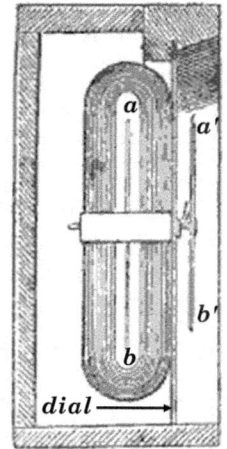

Fig. 5
Needle and
indicating hand

38. In order to govern the play of the needle, it is necessary that the agent at the station from which the signal is transmitted should have the power, 1st. To suspend and transmit the current at the receiving station; and 2nd. To change its direction upon the conducting wire. The former is necessary, to enable him to bring at all times the needle to its position of rest; and the latter, to deflect it to the right or to the left, according to the exigencies of the telegraphic communications.

The general principle on which these changes in the flow and direction of the current are effected, has been already explained. It is easy to imagine, that by very simple mechanism the movement of a lever or arm may make or break the contact of the conducting wires, so as to transmit or suspend the current at pleasure. Also by a simple motion of such an arm the contact makers, or their equivalents, may be moved from one side to the other, so as to reverse the current upon the wire to which the arm is directed.

If then an agent at the station be provided with any means of suspending or reversing the current which passes along the wire, he can at will bring a magnetic needle at the distant station to its position of rest, that is, to the vertical position, by suspending the current; or deflect it to the right, by causing the current to flow in one direction on the conducting wire; or to the left, by reversing the direction of the current.

The particular manner in which these several operations subserve to telegraphic purposes will be presently explained.

39. To explain the manner in which the electric current can impart temporary magnetism to soft iron, let us suppose a copper wire wrapped with silk, to prevent the metallic contact of contiguous convolutions, to be coiled round a rod of soft iron, bent into the form of a horse-shoe, as represented in Figure 6, care being taken, that in carrying

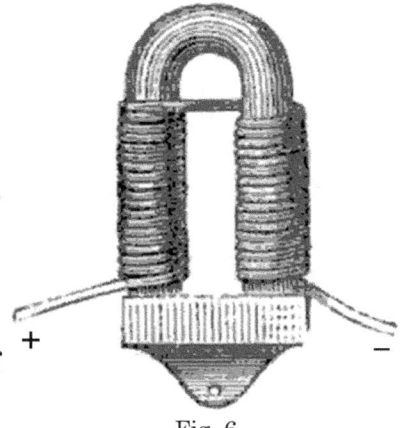

Fig. 6
An electro-magnet

the wire from one arm to the other, the direction of the convolutions shall be the same as if the coils had been continued round the bend.

So long as no electric current passes along the convolutions of the wire the horse-shoe will be free from magnetism. But if the ends of the wire, marked + and –, be put in connection with the poles of a voltaic battery, so that a current flows round its convolutions, the horse-shoe will instantly become a magnet, and will be so much the more powerful as the current is more intense, and the coils more multiplied.

If an armature loaded with a weight be presented to the ends of the horse-shoe while the current passes on the wire, it will adhere to them, and the weight, if not too great, will be supported.

In 1830 an electro-magnet of extraordinary power was constructed under the superintendence of M. Pouillet, at Paris. This apparatus, represented in Figure 7, consists of two horse-shoes, the legs of which are presented to each other, the bends being turned

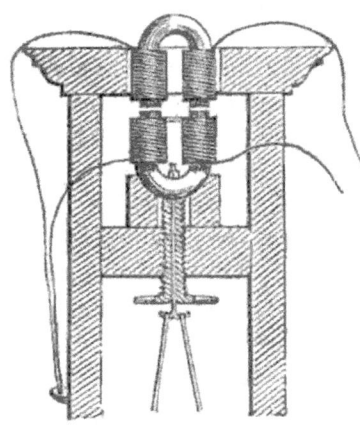

Fig. 7
A high-power electro-magnet

in contrary directions. The superior horse-shoe is fixed in the frame of the apparatus, the inferior being attached to a cross-piece which slides in vertical grooves formed in the sides of the frame. To this cross-piece a dish or plateau is suspended in which weights are placed, by the effect of which the attraction which unites the two horse-shoes is at length overcome. Each of the horse-shoes is wrapped with 10,000 feet of covered wire, and they are so arranged that the poles of contrary names shall be in contact. With a current of moderate intensity the apparatus is capable of supporting a weight of several tons.

It is found more convenient generally to construct electro-magnets of two straight bars of soft iron, united at one end by a straight bar transverse to them, and attached to them by screws, so that the form of the magnet ceases to be that of a horse-shoe, the end at which the legs are united being not curved but square. The conductor of the heliacal current is usually a copper wire of extreme tenuity.

40. In whatever form these magnets are constructed, the circumstance which in their telegraphic use is of most importance to notice, is that if proper conditions be observed in their preparation, their acquisition of the magnetic virtue upon the establishment of the current, and their loss of it upon the suspension of the current, are, for all practical purposes, instantaneous. The moment the extremities of the wire coiled round the horse-shoe are put into connection with the poles of the battery the horse-shoe becomes a magnet, and the moment the connection with the battery is broken it loses the magnetic virtue.

It has been already shown, that by means of very simple expedients, the current may be interrupted hundreds or even thousands of times in a second, being fully re-established in the intervals. The acquisition and loss of magnetism by the horse-shoe accompany these pulsations with the most perfect and absolute simultaneity. If the pulsations of the current be produced, at the rate of a thousand per second, the alternate presence and absence of the magnetic virtue in the horse-shoe will equally be produced at the rate of a thousand per second. Nor are these effects in any way modified by the distance of the place of interruption of the current from the magnet. Thus, pulsations of the current may be produced by an operator in London, and the simultaneous pulsations of the magnetism may take place at Vienna, provided only that the two places are connected by a continuous series of conducting wires.

41. It remains to show how these rapid pulsations of the magnetism of the bar can be rendered sensible, and how they may even be estimated and counted.

Let two straight rods of soft iron be surrounded by a succession of convolutions of covered wire, such as has been already described, and let the ends, *m*, *m'*, Figure 8, of these rods be connected by a straight bar of soft iron, attached to them by screws and nuts. Let the

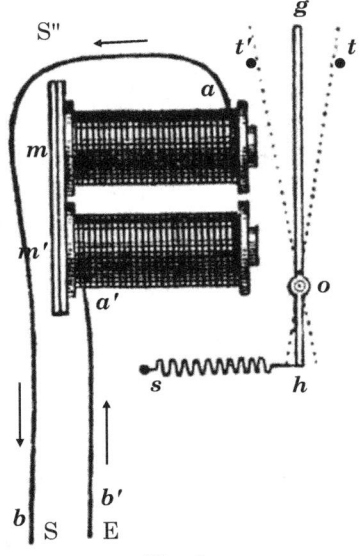

Fig. 8
Oscillations of an electro-
magnet armature

wire, *a b*, proceeding from a distant station, S, be put in metallic connection with the extremity of the wire coiled upon the rod, *m*, and let the wire, *a' b'*, connected with the extremity of the last convolution of the wire on the rod, *m'*, be put in metallic connection with the earth. If a current flow along *a b*, it will therefore circulate round the rods, *m* and *m'*, and will pass to the earth by the wire, *a' b'*. So long as this current flows, the rods will be magnetic, and they will lose their magnetism in the intervals of its suspension.

Let *g h* be a light iron bar, supported on a pivot, at *o*, on which it is capable of playing, so that its arm, *o g*, may move freely to the right or left. Let *t t'* be two stops, placed a small distance to the right and left of its extremity, *g*, so as to limit the range of its play. Let *s* be a spring attached to the extremity, *h*, by which that extremity will be constantly drawn to the left, and therefore the opposite extremity, *g*, thrown to the right against the stop, *t*. When the current is suspended, and the rods, *m m'*, divested of magnetism, the lever yielding to the action of the spring, *s*, the end, *g*, will rest against the stop, *t*. But when the current passes on the wire, the rods, *m m'*, becoming magnetic, will attract the arm, *o g*, of the lever, and this attraction exceeding the force of the spring, the arm, *o g*, will be drawn towards the electro-magnet, until it encounters the stop, *t'*, against which it will rest so long as the current continues to flow. But the moment the current is suspended, the bars, *m m'*, suddenly losing their magnetism, the lever, *o g*, is abandoned to the action of the spring, and it is again thrown back upon the stop, *t*, where it rests until the current is re-established.

Let us suppose that an agent at the station, S, to which the wire, *a b*, extends, and which may be at any distance, 500 miles for example, from S", is supplied with any of the means which have been explained, by which he can at will control the pulsations of the current. When he causes the current to flow, he imparts magnetism to the bars, *m m'*, and throws the lever, (or armature) *o g*, against the stop, *t'*. When he suspends the current he deprives the bars, *m m'*, of their magnetism, and leaves the lever, *o g*, to the action of the spring, *s*, by which it is thrown against the stop, *t*.

It appears, therefore, that with each pulsation which the current receives from the agent at S, the armature, *o g*, at S", 500 miles distant from him, will perform a vibration between the stops, *t* and *t'*. As the transmission and suspension of the current, and also the acquisition and loss of the magnetic power, by the rods, *m m'*, are both instantaneous, there is no practical limit to the velocity of the pulsations of the current and those of the magnetism alternately acquired and lost by the rods, *m m'*. The oscillations of the armature, *o g*, produced by these pulsations are limited, however, by its weight, the force of the spring, and the distance between the stops, *t* and *t'*. The greater the weight of the lever, the force of the spring, and the distance between the stops, the slower will be the motion of the lever from *t* to *t'*, produced by a current of given intensity. The greater the weight of the armature, and the distance between the stops, and the less the force of the spring, the slower will be the motion from *t'* to *t*.

The stop, *t'*, is so placed as to prevent the absolute contact of the arm of the armature with the electro-magnet, but to allow it to approach the latter very closely. Absolute contact is to be avoided, because it is found that in that case the arm adheres to the magnet with a certain force after the current ceases to flow, but so long as absolute contact is prevented, it is immediately brought back by the spring, *s*, when the current is suspended.

42. It is evident, therefore, that the limit of the possible celerity of vibration to be imparted to the armature, *o g*, by the pulsations of the current will depend on the nice adjustment of the weight and play of the armature, and the force of the spring, *s*.

The velocity of oscillation, however, which can in this way be given to the armature, is such as can scarcely be credited without actually witnessing its effects. When that velocity does not exceed a certain limit the oscillations may be registered and counted, by causing the armature to give motion to the anchor of an escapement, connected with a train of wheel-work, by which a hand or index, moving on a graduated dial, is governed. But these oscillations are susceptible of velocities so great that it would be difficult to apply this expedient for counting them. M. Gustave Froment, of Paris, suggested and applied to this purpose with complete success, a method of ascertaining the velocity depending on the laws which govern the vibrations of musical strings.

43. It is well known that the pitch of any musical note is the consequence of the rate of vibration of the string by which it is produced, and that the more rapid the vibration the higher the note will be in the musical scale, and the slower the vibration the lower it will be. Thus the string of a pianoforte which produces

the bass note 𝄢 vibrates 132 times in a second, that which

produces the note 𝄢 vibrates 66 times in a second, and that

which produces the note 𝄞 vibrates 264 times per second.

On a seven-octave pianoforte the highest note in the treble is

three octaves above 𝄞, and the lowest note in the bass is four

octaves below it. The number of complete vibrations corresponding to the former must be 3520; and the number of vibrations per second corresponding to the latter is 27½.

If, therefore, the lever, *o g*, have any rate of vibration more rapid than 27½ vibrations per second, and less rapid than 3520 per second, it will produce by its motion some definite musical sound, and if the note formed upon a pianoforte, which is in unison with it, be found, the rate of vibration of the string producing that note will be the same as that of the lever.

When it is stated that the vibrations imparted by the pulsations of the current to levers, mounted in the manner here described, have produced musical notes nearly two octaves higher than the highest note on a seven-octave piano, tuned to concert pitch, it may be conceived in how rapid a manner the transmission and suspension of the electric current, the acquisition and loss of magnetism in the soft iron rods, and the consequent oscillation of the lever, upon which these rods act, take place. The string which produces the highest note, on such a piano, vibrates 3520 times per second. A string which would produce a note an octave higher would vibrate 7040 times per second, and one which would produce a note two octaves higher would vibrate 14,080 times per second.

It may, therefore, be stated, that by the marvellously subtle action of the electric current, the motion of a pendulum is produced, by which a single second of time is divided into from twelve to fourteen thousand equal parts!

44. It has been already shown how the motion of clockwork may be applied to control and regulate the pulsations of the electric current. We shall now show how, on the other hand, the pulsations of the current may be made to govern the motion of wheelwork. This expedient must be regarded with the more interest inasmuch as it has been applied with the greatest effect in several of the varieties of electric telegraph, which have been proposed or brought into operation.

45. If we suppose the armature *g h*, Figure 8, to be put into connection with the anchor of the escapement wheel of a system of clockwork, it will be easy to see how that clockwork can be regulated by the pulsations of the electric current.

In Figure 9, *w w'* is the escapement wheel which is constantly

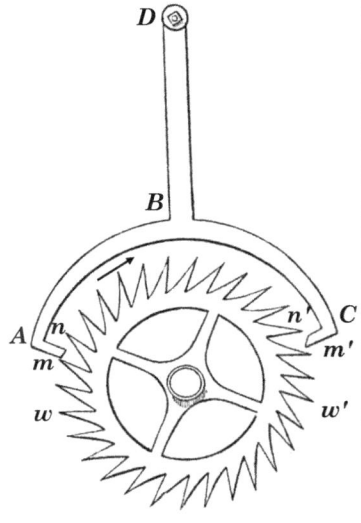

Fig. 9
Clockwork escapement wheel

impelled by the force of a descending weight or mainspring in the direction of the arrows. The anchor, **A B C**, of the escapement is connected with an axis **D**, by the straight rod **B D**. This rod **B D** may be either the vibrating arm of a lever, such as **g h**, Figure 8, kept in oscillation by the current acting on an electro-magnet, or it may be connected with such a lever in any convenient manner, so as to oscillate simultaneously with it, and to have the extent of play necessary for the action of the pallets **A** and **C** of the anchor on the teeth of the escapement wheel.

When the anchor is not in a state of oscillation, a tooth of the wheel will rest upon one of its pallets, and the wheel and clockwork connected with it will be stopped. When the anchor moves from left to right, the tooth of the wheel, which was previously stopped by the upper surface **n'** of the pallet **C**, is allowed to *escape*, and in obedience to the power of the spring or weight, which moves the clockwork, it advances towards **m'**. Meanwhile the pallet **A** enters the space between two teeth of the wheel, one of which coming against its lower surface, it stops its motion. When the anchor moves back from right to left, the pallet **C** comes under the next tooth of the wheel. In this manner every movement of the anchor to the right lets a tooth, which was stopped by the pallet **C**, advance, and afterwards the pallet **A** stops the advance of another tooth, while every movement to the left lets the tooth stopped by **A** advance, and afterwards the pallet **C** stops the next tooth which advances on that side.

Thus each complete oscillation of the anchor, consisting of a motion to the right and a motion to the left, lets one tooth of the escapement wheel, and no more, pass.

Now if we suppose the pulsations of the current to impart to the anchor by the intervention of the electro-magnet and its appendages a motion of vibration, a tooth of the escapement wheel, and no more than one tooth, will pass the anchor for each pulsation of the current. If the current be suspended the movement of the escapement wheel and the clockwork connected with it will be also suspended, and when the pulsation of the current recommences, the oscillations of the anchor, and consequently the motion of the escapement wheel, and the clockwork connected with it, will also recommence.

46. If the pulsations of the current be regulated (as they may be according to what has been already explained), by the pendulum of a clock at any station, the motion of the anchor of the escapement, established at any other station to which the current is transmitted, will be synchronous with that of the pendulum of the clock which governs the pulsations of the current; and thus a regular motion may be imparted by one clock to another, provided that between them there be established a conductor; and the pendulum of the one clock regulates the pulsations of the current, which govern the movement of the anchor of the escapement of the other.

47. If the extremity of the lever, *o g*, Figure 8, carry a pencil, which presses upon paper, when the lever is drawn towards the electro-magnet, and if at the same time the paper is moved under the pencil with a uniform motion, a line will be traced upon the paper by the pencil, the length of which will be proportionate to that of the interval during which the lever *o g* is held in contact with the stop *t'*. Now as the agent at S can regulate this interval at will, by controlling the flow of the electric current, making that current act for a short interval if he desire to make a short line upon the paper, for a long interval if he desire to make a long line, and for an instant if he desire to make merely a dot, it will be understood how he can at will mark a sheet of paper at S", 500 miles distant, with any desired succession of lines of various lengths or of dots, and how he may combine these in any way he may find suitable to his purpose.

We have here supposed the pencil attached to the end of the lever to be alternately pressed against the paper, and withdrawn from it by the motion of the lever. If, however, the paper be so placed that the lever shall oscillate parallel to it, the pencil presented to the paper will remain permanently in contact with it, and will trace upon the paper a line alternately right and left, whose length will be equal to the play of the end *g* of the lever, to which the pencil is attached. If while this takes place the paper be moved under the pencil in a direction at right angles to the line of its play, the pencil will trace upon the paper a zigzag line, the form of which will depend on the relation between the motion of the paper and that of the pencil. When the current is in this case suspended, the paper being moved under the pencil at rest, a straight line will be traced upon it.

Thus the paper will be marked either with a zigzag line or a straight line, according as the current is transmitted or suspended.

If the current be alternately transmitted and suspended during intervals of unequal length, at the will of the agent, at S, the paper at S" will be marked by a line alternately zigzag and straight, the length of the zigzag and straight parts being varied at the will of the operator at S.

How these subserve to telegraphic purposes will be presently more fully explained.

48. In the same manner, if a toothed wheel, moved by the agent at S, produce a pulsation of the current by the passage of each successive tooth, these pulsations will produce simultaneous oscillations of the lever *o g*, at the station S", and if these oscillations act upon the anchor of an escapement wheel attached to clockwork at S", that wheel will be advanced in its revolution, tooth for tooth, with the wheel at S, and if each of these wheels govern the motions of hands upon dial plates, like the hands of a clock, the hand of the dial at S" will have the same motion exactly as the hand on the dial at S, so that if at the commencement of the motion both hands point to the same figure or letter of the dial, they will continue, moving together, to point always to the same figures or letters.

Thus if the operator at S desire to direct the hand on the dial at S", to the hour of 3 or 5, he will only have to turn the hand upon the dial, at his own station, to the one or the other of these hours.

It will presently, also, be apparent how important this is in the art of electro-telegraphy.

49. If the lever *o g*, Figure 8, be connected with the tongue of an alarum-bell, so that when *o g* is put into vibration the bell will ring, and will continue to ring so long as the vibration is continued, it is evident that the operator at S can, at will, ring a bell at S″, by producing pulsations of the current in any of the ways already described.

An operator at S″ may in like manner ring a bell at S.

By this mutual power of ringing bells, each operator can call the attention of the other, when he is about to transmit a despatch, and the other by ringing in answer can signify that he is prepared to receive the despatch, as already stated.

50. If the lever *o g* were in connection with the lock or other mechanism, by which the powder charging a cannon is fired, the operator at T could at will discharge a cannon at R, no matter what may be the distance of R from T.

51. It will be observed that when a bell is rung, or any similar signal produced at the station S″, by means of an electric current transmitted from a distant station, S, it is not directly the force of the current which acts upon the object by which the signal is made. The current is only indirectly engaged, producing the result by liberating the mechanism which makes the signal and leaving the force which moves it free to act. Thus in the most usual case of a bell, it is acted upon while it rings, not by the current, but by the force of a mainspring or descending weight, transmitted to the hammer or tongue in the same manner exactly as that in which the force of a mainspring or weight of a clock is transmitted to the striking apparatus. The current does nothing more than disengage a catch by which the motion of the wheelwork acted on by the mainspring or weight is arrested. The catch once disengaged, the action of the current on the bell ceases, and the ringing is continued by the action of the mainspring or weight, and it may in like manner be stopped by the current again throwing the catch between the teeth of one of the wheels.

It will, therefore, be apparent that since the force which impels the bell is independent of the current, a bell of any desired magnitude may be acted upon by a hammer of any desired weight, without requiring any more force from the current than that which is sufficient to enable the electro-magnet to disengage the catch by which the mechanism of the bell is arrested.

52. Although the bell mechanism used for telegraphs differs in nothing which is essential from that of a common alarum clock, it may not be without interest to show one of the varieties of mechanism in practical use.

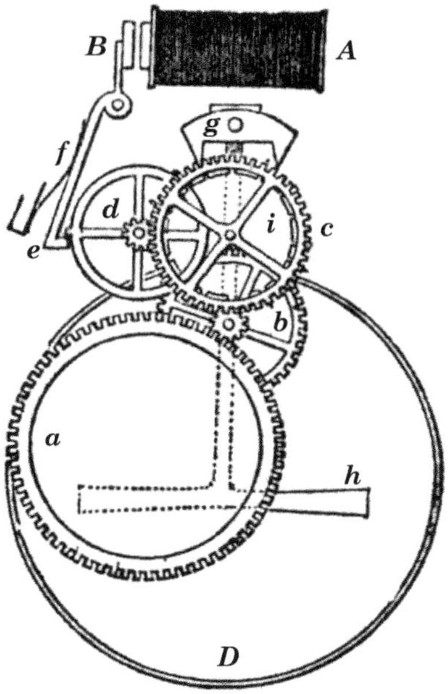

In Figure 10 is given a view of the bell mechanism, as used on the telegraphic line of the South-Eastern Railway Company.†

A is the electro-magnet.

B its armature.

Fig. 10
Alarum bell mechanism

B e a lever attached at the upper end to the armature, and having at the lower end a catch, *e*, which when the armature is not attracted towards the magnet is pressed by a spring, *f*.

d a wheel having a tooth in which the catch *e* is engaged by the pressure of the spring *f*, when the armature *B* is not attracted towards the magnet, but which is liberated from the catch *e*, when the armature *B* is drawn towards the magnet.

a a cylindrical box containing a strong mainspring, by which the train of wheelwork is kept in motion so long as the catch *e* is not engaged in the tooth of the wheel *d*.

The actual contact of the armature *B* with the poles of the electro-magnet is prevented by two small ivory knobs screwed into the surface which is presented to the magnet. The play of the armature *B* is so limited that the catch *e* shall be just disengaged from the tooth of the wheel *d* when the ivory knobs come into contact with the poles of the magnet.

† Elect. Tel. Manip., p. 23.

When the wheel-work is liberated by the magnet withdrawing the catch *e* from the wheel *d*, the mainspring in the cylindrical box *a* causes the toothed wheel attached to the box to revolve. This wheel drives a pinion on the axle of the wheel *b*; the wheel *b* drives a pinion on the axle of the wheel *c*; the teeth of the wheel *c* are engaged with those of a pinion on the wheel *d*. The movement of the train is stopped when the catch *e* falls under the tooth of the wheel *d*. The wheel *i*, which is engaged in the anchor of the escapement *g*, is fixed upon the axle of the wheel *c*, turns with the latter, and thus gives an oscillating motion to the anchor, which is imparted to the hammer *h* of the bell *D*. The bell is therefore acted upon by the hammer so long as the magnet *A* keeps the catch *e* from falling under the tooth of the wheel *d*.

53. Since the magnitude, loudness, or pitch of the bell is independent of the force of the current, the telegraphic offices are provided with various bells for special purposes.

Sometimes a special wire is appropriated to the bell which is acted upon by a special current. In other cases the regular current intended to work the telegraph is diverted to the bell apparatus by the commutator. In other cases, again, the object is accomplished by cutting the bell off from circuit on commencing message transmission.

54. Having explained the form and construction of electro-magnets, we are prepared to show the manner in which an electric current may be produced by the mere action of magnets without any intervention of a voltaic battery.

The electricity thus produced has been called MAGNETO-ELECTRICITY.

55. Let a silk or cotton covered wire be coiled heliacally on a roller or bobbin having a hollow core of sufficient magnitude to allow a cylindrical bar to be passed into it. Let the covered wire be coiled constantly in the same direction, beginning from *A B* (Figure 11), and terminating at *C D*. Let the extremities *m n* of this wire be joined to those of a wire *m o n* of any required length, stretched to any required distance. Now let the north pole N of a magnet S N be suddenly passed into the core of the bobbin. An electric current will then be transmitted on the wire *m o n*, the

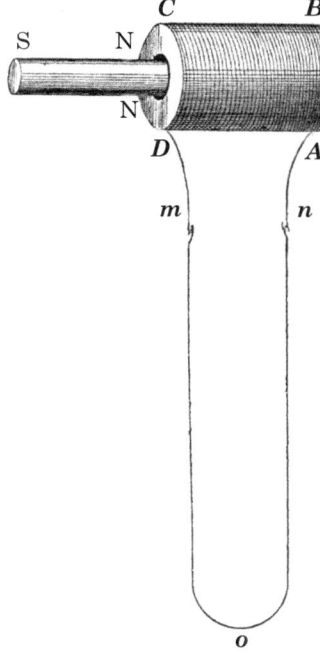

Fig. 11
Magneto electricity

presence of which may be rendered manifest by a galvanometer. This current, however, will be only momentary, being manifested only at the moment the pole of the magnet enters the core of the bobbin. It ceases immediately after that entrance.

Now if the magnetic bar after entering be as suddenly withdrawn, another current will be produced upon the wire *m o n*, which will also be only momentary, but its direction on the wire will be contrary to that produced by the entrance of the magnetic pole.

Thus if upon the entrance of the pole N a current is produced, running in the direction *m o n*, the withdrawal of the pole N will produce a current running from *n o m*.

If the south pole S be passed into the core and withdrawn, momentary currents will in like manner be produced, but they will have contrary directions.

If the wire *m o* terminated at *o*, no matter what may be the distance of *o* from *m*, were put at *o* in metallic communication with the earth, or with a plate or other mass of metal buried in the earth, and if the extremity *n* of the wire of the coil were put in metallic connection with the earth in the same manner at *n*, the transmission of the instantaneous currents would take place exactly in the same manner as above described, because in that case the earth would play the part of a conductor between the end of the wire *m o* at *o*, and the end of the coil wire *n*.

But if the metallic continuity either of the wire *m o n*, in case it extended from *m* to *n*, or of *m o* if it were as described above in connection with the earth at *o*, were anywhere broken, no current would be produced by the entrance or withdrawal of the magnet. It is therefore essential to the production of these phenomena that the extremities *m* and *n* of the coil wire shall be in electric communication with each other, by being united either with a continuous metallic connection, or by means of the earth in the manner already described.

The property in virtue of which soft iron acquires magnetic properties, when the poles of a permanent magnet are brought into proximity with it, supplies a very convenient method of exhibiting the play of the phenomena of momentary currents above described.

56. Let *S O N* (Figure 12), be a powerful permanent horseshoe magnet, having its poles *S*, *N*, presented to and in close proximity with a similar horse-shoe *a b* of soft iron, wrapped

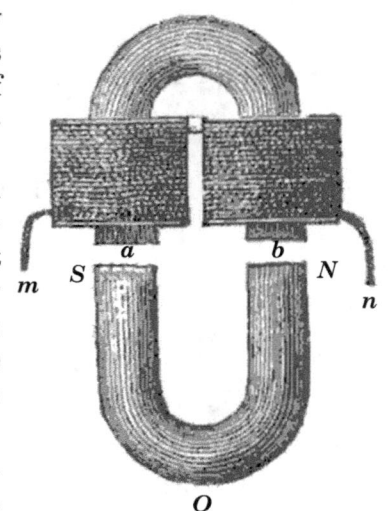

Fig. 12
Phenomena of momentary currents

with convolutions of covered wire in the manner already described. Let the extremities *m* and *n* of the coil be supposed to be placed in connection with two wires, which may be extended to any distances, and whose extremities are in metallic communication with the earth in the manner already explained. When the poles *S* and *N* are brought into proximity with the ends *a* and *b* of the horseshoe *a b*, the latter will, by the inductive action of the magnet *S O N*, acquire magnetic polarity, the end *a*, near the south pole *S*, having northern, and the end *b*, near the north pole *N*, having southern polarity. This magnetic polarity, however, of *a b* will only continue so long as the poles *S* and *N* of the permanent magnet are kept near to *a* and *b*. If they be removed, that instant the polarity of *a b* will cease. If the poles be reversed, *N* being presented to *a*, and *S* to *b*, then *a* will acquire south, and *b* north polarity.

It appears, therefore, that by presenting the poles of the magnet *N O S* to the horseshoe, the same effect is produced as if the poles of a magnet were suddenly passed into the axis of the coil, and by withdrawing the poles *N* and *S* from *a* and *b*, the same effect is produced on the coil as if the poles of the magnet which had been passed along the axis were suddenly withdrawn.

57. The momentary currents in the one direction or in the other will, therefore, be produced upon the wire connected with the extremities of the coil, such as have already been described, each time the poles *N* and *S* are presented to and withdrawn from the ends, *a* and *b*, of the horseshoe of soft iron. If the magnet *N O S* were mounted so as to revolve upon an axis passing through the centre of its bend, and therefore midway between its legs, its poles might be made to pass the ends of the horse shoe, the latter being stationary. During each revolution of the magnet *N O S* the polarity imparted to the horse-shoe would be reversed.

When the pole *N* approaches *b*, and consequently *S* approaches *a*, south polarity will be imparted to *b*, and north polarity to *a*; and when *N* passes *a*, and consequently *S* passes *b*, south polarity will be imparted to *a*, and north polarity to *b*.

The momentary currents produced by these changes of magnetism in *a* and *b* will be easily understood by what has been explained. When *N* approaches *b*, and *S* approaches *a*, the commencement of south polarity in *b*, and north polarity in *a*, will both impart to the wire a current in the same direction, because the coils of the spiral as presented to *S* will be the reverse of those presented to *N*. When *N* departs from *b*, and *S* from *a*, the cessation of south polarity in *b*, and of north polarity in *a*, will impart currents in the same direction to the wire, but this direction will be opposite to that of the former currents.

When *N* approaches *a*, and consequently S approaches *b*, currents will be imparted to the wire whose direction will be the same as that of those produced by the departure of *N* from *b*, and of *S* from *a*. When *N* departs from *a*, and *S* from *b*, currents will be produced in the same direction as when *N* approaches *b* and *S* approaches *a*.

If the direction of the currents produced when *N* approaches *b*, and *S* approaches *a*, be indicated by an arrow directed to the right, and that of those produced when *N* departs from *b*, and *S* from *a*, by an arrow directed to the left, the changes of direction which take place in each revolution of the magnet *N O S*, will be such as are indicated in Figure 13, where *b* and *a* represent the ends of the horseshoe **b a**; **N** the position of the pole in approaching, and **N'** in departing from **b**, and **N''** its position in approaching, and **N'''** in departing from **a**. The arrows directed to the right express the direction of the two currents which are produced upon the conducting wire, while **N** makes the half revolution **N''' M' N**; and the arrows directed to the left express the direction of the two currents produced, while **N** makes the half revolution **N' M N''**.

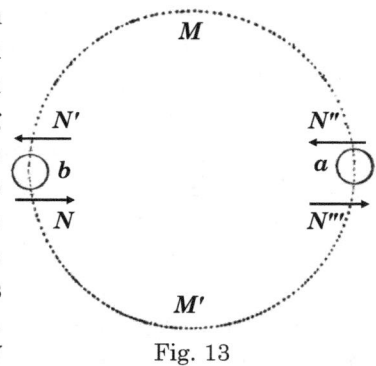

Fig. 13
Changes of current direction

Thus it appears that in each revolution of the magnet **N O S**, four momentary currents are produced in the wire, two in one direction during one semi-revolution, and two in the contrary direction during the other semi-revolution. In the intervals between these momentary currents there is a suspension of voltaic action.

58. The electric currents may be instantaneously suspended, re-established, and reversed in their direction by means of a reversing wheel or commutator to change the connections. By such an expedient properly adapted, it is easy to understand that by suspending the currents in one of the two contrary directions, while the other is allowed to pass, an intermitting current always running in the same direction may be obtained. Or if the commutator be so adapted that while the momentary currents in one direction are allowed to run without interruption, those in the other direction shall be reversed, we shall then have in each revolution four momentary currents flowing in a common direction. The current thus produced will be intermitting, that is, it will pass upon the wire by a succession of pulsations or intervals of transmission and suspension, but since in each revolution of the magnet there are two pulsations,—that is, two intervals of transmission and two of suspension,—and since the rotation of the magnet may be made with any desired rapidity, it follows that the pulsations will succeed each other with such celerity, and the intervals of suspension will be so brief, that for all practical purposes the current will be continuous.

59. Such are the principles on which is founded the construction of magneto-electric machines, one form of which is represented in Plate II (p. 26). The purpose of this apparatus is to produce by magnetic induction an intermitting current constantly in the same direction, and to contrive means by which the intervals of intermission shall succeed each other so rapidly that the current shall have practically all the effects of a current absolutely continuous.

A powerful compound horseshoe magnet, *A*, is firmly attached by bolts and screws upon a horizontal bed, beyond the edge of which its poles *a* and *b* extend. Under these is fixed an electro-magnet *X Y*, with its legs vertical, and mounted so as to revolve upon a vertical axis. The covered wire is coiled in great quantity on the legs *X Y*, the direction of the coils being reversed in passing from one leg to the other.

The two extremities of the wire proceeding from the legs *X* and *Y* are pressed by springs against the surfaces of two rollers, *c* and *d*, fixed upon the axis of the electro-magnet. These rollers themselves are in metallic connection with a pair of handles, *P* and *N*, to which the current evolved in the wire of the electro-magnet *X Y* will thus be conducted.

If the electro-magnet *X Y* be now put in rotation by the handle *m*, the handles *P* and *N* being connected by any continuous conductor, a system of intermitting and alternately contrary currents will be produced in the wire and in the conductor by which the handles *P* and *N* are connected. But if the rollers *c* and *d* are so contrived that the contact of the ends of the wire with them shall be only maintained during a semi-revolution in which the intermitting currents have a common direction, or so that the direction during the other semi-revolution shall be reversed, then the current transmitted through the conductor connecting the handles *P* and *N* will be intermitting, but not contrary; and by increasing the velocity of rotation of the electro-magnet *X Y*, the intervals of intermission may be made to succeed each other with indefinite celerity, and the current will thus acquire all the character of a continuous current.

The forms of commutators by which the rollers *c* and *d* are made to break the contact, and re-establish it with the necessary regularity and certainty, or to reverse it during the alternate semi-revolutions, are various.

All the usual effects of voltaic currents may be produced with this apparatus. If the handles *P* and *N* be held in the hands, the arms and body become the conductor through which the current passes from *P* to *N*. If *X Y* be made to revolve, shocks are felt, which become insupportable when the current has a certain intensity.

If it be desired to give local shocks to certain parts of the body, the hands of the operator, protected by non-conducting gloves, direct the knobs at the ends of the handles to the parts of the body between which it is desired to produce the voltaic shock.

For telegraphic purposes it will be sufficient to place the line wire in connection with one of the handles *P* or *N*, while the other handle is in connection with the earth. A current will then be transmitted on the line wire which will be intermitting, but which may be rendered continuous by a combination of magneto-electric machines.

60. It remains, in fine, to show how the chemical properties of the electric current can be made to supply the means of transmitting signals between two distant stations.

When a current of adequate intensity is made to pass through certain chemical compounds, it is found that these are decomposed, one of their constituents being carried away in the direction of the current, and the other in the contrary direction.

One of the most striking examples of the application of this principle is presented in the case of water, which, as is well known, is a compound of the gases called oxygen and hydrogen.

Let us suppose that a series of cups, *o h* (Figure 14), containing water, are placed so that an electric current shall pass successively through them, commencing at the wire *P*, and passing at *o* into the first cup; thence through the water to *h*, and from *h* along the wire *I* to *o* in the second cup; thence in like manner through the water to *h*, and then along the wire *I'*, and so on to *N*, the wire *P* being supposed to be connected with the positive pole of a battery, and the wire *N* with its negative pole. The current will therefore flow from **P** to **N**, passing through the water in each of the cups. Under such circumstances the

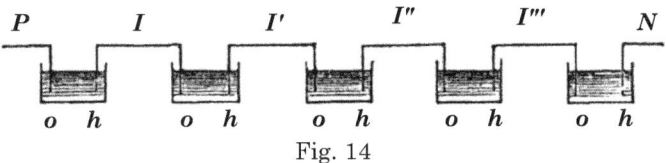

Fig. 14

Decomposition of water by an electric current

water will be gradually decomposed in each of the cups, the particles of oxygen moving against the course of the current, and those of hydrogen moving with it, the former are evolved at the points **o**, and the latter at the points **h**.

To show how this property of the current may be made to produce visible marks or signs, let us suppose a sheet of paper, wetted with an acidulated solution of ferro-prussiate of potash, to be laid upon a plate of metal, and let the point of a metallic style be applied to it, so as to press it gently against the metallic plate without piercing it. Let the style be now put in metallic connection with the wire which leads to the positive pole of a voltaic battery, and let the metallic plate upon which the paper is laid be put in connection with the wire which leads to the negative pole. The current will, therefore, flow from the style through the moistened paper to the metallic plate, and it will decompose the prussiate, one of the constituents of which, deposited on the paper, will mark it with a bluish spot.

If the paper be moved under the style while the current flows, this decomposition being continued under the point of the style, a bluish line will be traced upon the paper.

If while the paper is thus moved uniformly under the style, the current is permitted to flow only during intervals long or short, the paper will be marked by lines long or short, according to the intervals during which the current flows; and, since no decomposition takes place during the suspension of the current, the paper then passes under the style without receiving any mark. If the current be permitted to flow only for an instant, the paper will be marked by a dot. The long or short lines and dots, thus traced upon the paper, will be separated one from another by spaces more or less wide according to the lengths of the intervals of suspension of the current.

PLATE III

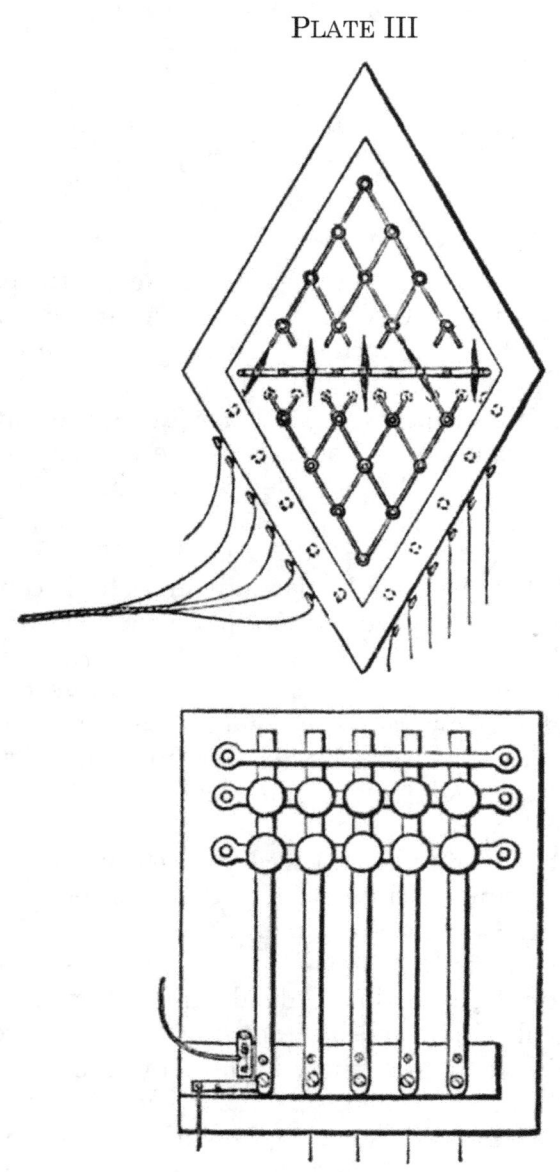

Cooke and Wheatstone's five-needle telegraph, 1837
(*top*: the indicating needles, *bottom*: operating keys)

3

THE HISTORY OF THE TELEGRAPH

61. HAVING thus described the theory of electricity, we will now
proceed to treat of its application to telegraphic purposes.

The conveyance of intelligence by means of preconcerted signals
has been practised with more or less success from the most ancient
times. The use of beacons by night, and columns of smoke by day,
as a mode of communicating events, is referred to in the sacred
writings; and we have had a comparatively recent instance of the
continued existence of a rude method of passing a signal from
hand to hand, in the circulation of the cakes employed by the natives
of India in arranging the outbreak of 1857.

62. These signals were gradually improved and supplemented by
mechanical contrivances until shortly before the French revolution.
Claude Chappé, a young student in the College at Angers, contrived
a telegraph by means of which he could confer with his two brothers,
who were placed at a school about a mile and a half distant from
the town. The apparatus consisted of a wooden beam rotating upon a centre, hearing at its extremities two arms capable of being
moved around their axes to any required angle.

This simple machine was capable of producing 192 separate
signals, and was afterwards adopted by the French Government,
stations being established throughout the country. Lines of semaphore telegraph were soon after erected in England on a somewhat similar plan, and a comparatively high rate of speed was

attained by them. The frequent liability to interruption from misty weather was however, especially in this climate, a great defect, and was not uncommonly the cause of much confusion.

Thus, on an occasion when, during the Peninsular war, the admiral at Plymouth had an important message to transmit to Whitehall, he was only able to forward part of it at first, a thick fog gathering over a portion of the line of stations, and interrupting the communication. Great distress and anxiety resulted in London, for the first part of the message transmitted was "Wellington defeated;" the remainder of the message which came next day, "the French at Salamanca," changed the metropolitan sorrow into gladness.

It would be foreign to the object of the present work to enter at any length into the various ingenious mechanical contrivances for communicating visual signals by day and night.

63. Our readers are probably aware that it was not until the middle of last century that the discovery of the power of generating and collecting electricity derived from the friction of glass and other substances, drew the attention of men of science to the study of phenomena, which had been observed to a trifling extent by the philosophers of ancient times. It was not long before it was ascertained that the effect could be communicated along insulated conducting substances to a considerable distance.

At the Carthusian Convent, in Paris, the whole community stood in a line holding pieces of iron wire to a length of some 2000 yards, and on discharging an electrical jar at the end of the line the whole company gave a sudden spring at the same instant, all feeling the shock together.

In England also Sir William Watson, a distinguished man of science, carried out experiments on a much larger scale, and by means of a wire erected at Shooter's Hill for a length of 12,000 feet, showed that the transmission of electricity was to all perception instantaneous.

64. It might be supposed that this having been established, and semaphore apparatus being in existence by which signals were daily conveyed from place to place, that some instrument would very soon have been constructed by which electricity could be applied to communicating intelligence. This was in fact tried in various ways by Lesage, Lomond, Ronalds, and others, several lines of more or less length being constructed in different countries. The difficulty, however, of sufficiently insulating the wires to retain electricity of high tension and small quantity was so great, that in no instance was a success attained which bore any practical fruit.

The general theory of electro-magnetic phenomena, so far as they bear upon telegraphy, has been described in the previous chapters. It is, however, worthy of remark that efforts were made for many years to apply the volatile species of electricity derived from friction to the production of signals, before the great discoveries of Galvani, Volta, Oërsted, and Ampere had furnished us with an agent sufficiently tractable for practical application.

65. During the early part of the present century the investigation of the attributes and phenomena of voltaic electricity (or that derived from the chemical decomposition of water already described) and its connection with magnetic action, gave rise to various suggestions and experiments by Soemmering, Schilling, Fechner, Ritchie, Steinheil, Gauss and Weber, and others, for the adaptation of this new power to signalling purposes.

66. The apparatus constructed by the last-named experimentalists upwards of thirty years ago deserves more than a passing notice, both from the remarkable ingenuity shown in multiplying the effect of the signal by the angle of reflection; and from the circumstance that it corresponds entirely in principle, and very nearly in detail, with the apparatus actually used for receiving signals through the Atlantic Cables by a reflecting galvanometer needle.

Gauss and Weber's telegraph consisted of a magnetic needle moved by currents developed by induction from a permanent magnet, the signals being shown by means of a small mirror attached to the needle, in order to increase the effect of its movements to the operator, who read the deflections by aid of a magnifier at a short distance.

This apparatus is shown in Figure 15.

a, is a small mirror with a counter-balance attached to the axle of a magnet suspended by a thread and actuated by a coil *b*, *b*. The scale *c*, shown separately in front view at *c'*, was fixed to the

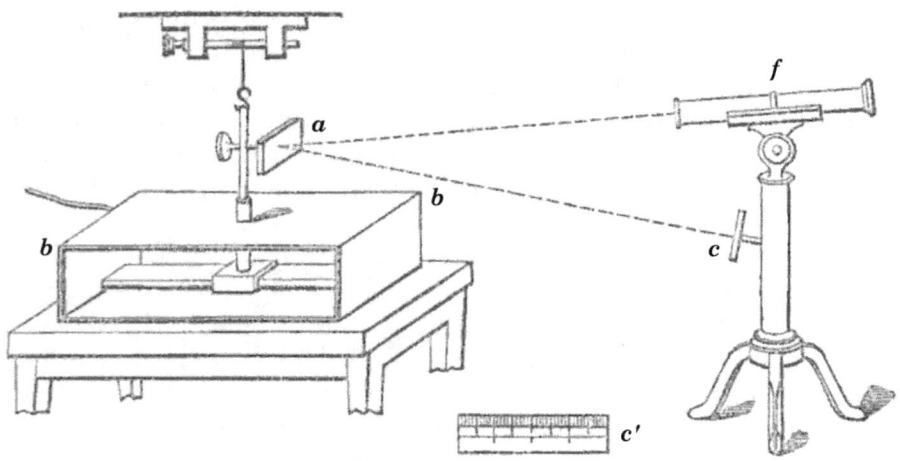

Fig. 15
Gauss and Weber's telegraphic instrument

stand of the observing telescope *f*, so that it was reflected back to the eye of the observer. By this means a very small movement of the magnet could be read off. This telegraph was actually put in operation between the Cabinet of Natural Philosophy and the Observatory at Gottingen, about a mile and a half apart, by means of a copper wire suspended in the air.

67. It became a common experiment at the lecture hall table, at classes and philosophical meetings, to exhibit apparatus in which magnetic needles were deflected at a short distance from the exciting power. It happened that early in 1836 Mr. W. F. Cooke, who had previously held a commission in the Indian army, happened to see one of these experiments exhibited at Heidelberg. His mind was deeply impressed with the subject, for he saw at once that something more was to be done with this subtle power, than merely demonstrating that a magnetised needle could be deflected in one room by a battery in another; and from that day he devoted himself to the practical realization of the electric telegraph.

So rapidly did execution follow thought, that within three weeks he had constructed a reciprocal telegraph, by which letters could be interchanged between two distant stations, by the movement of three manipulating keys, with corresponding magnetic indices.

This apparatus is shown in Figure 16. Six wires were used.

Fig. 16
Cooke's first telegraph, 1836

This instrument possessed the important feature of reciprocality for receiving and indicating the signals; and for transmitting the currents from either end. The signals were shown by the turning of discs attached to the axles of three magnetic needles, when one or other of the needles was deflected by a current.

The indicating apparatus at the two termini being part of the circuit, whether the current was transmitted from one end or the other, every signal was exhibited at both places simultaneously, and the sender could perceive and correct any error of his manipulation, while the recipient of the message could immediately check him should any signal be misunderstood. Mr. Cooke shortly after devised a simple means of attracting the attention of a distant station by sounding an alarum, which has continued in general use ever since without material change in form or principle. It consisted of a train of wheels driven by a spring, and restrained by a detent, to which was fixed a piece of iron opposite the poles of an electro-magnet. On the current passing the clockwork mechanism was released, bringing a hammer into contact with a bell.

This alarum is shown at Figure 17.

68. After constructing various instruments (some of which in January, 1837, were brought before the Manchester and Liverpool Railway, with the object of applying them for signalling through the tunnel at Edge Hill), Mr. Cooke became acquainted with Professor Wheatstone, who had been engaged for some time upon a series of experiments regarding the velocity of electricity, and had also devoted some attention to the subject of telegraphing;

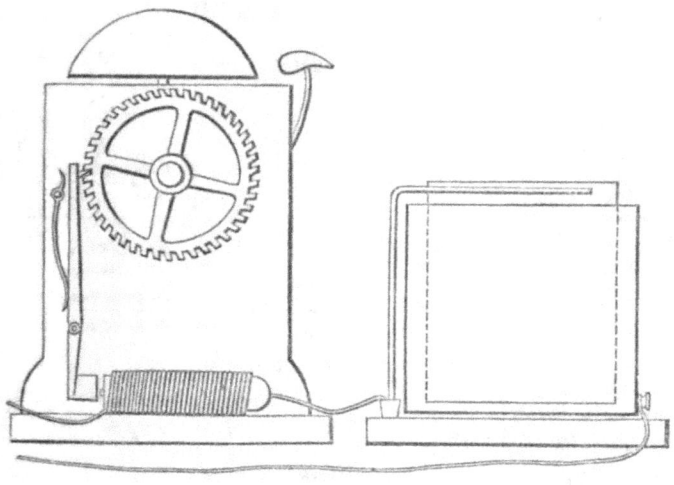
Fig. 17
Cooke's alarum

their discussions resulted in a combination of plans, and in the summer of the same year they obtained a patent for their united inventions, the principal features of which are shown at Plate III (p. 50); one part of which represents the sending apparatus, and the other the receiving dial with five needles.

The alphabet was indicated by the concurrent pointing of any two of the needles, the small circles shown in the figure being inscribed with the various letters.

The commutator or key-board for changing the currents consisted simply of five springs united together by a metallic bridge; finger studs being so placed that one battery pole or the other could be readily connected to each of the five conducting wires.

Although this apparatus was the parent of the simpler needle instruments, which are still largely used for railway and other purposes, it possesses only an historical interest; and together with various other early meritorious inventions of Messrs. Cooke and Wheatstone need not be further described.

Before passing on to the telegraphs actually in use, it should be mentioned that Professor Morse of America, (whose system was put into practical shape at a later period,) has shown that the germ of the recording apparatus which has since been so generally adopted, was the subject of some experiments by him in America, at a time slightly anterior to the telegraph of Messrs. Cooke and Wheatstone.

In Bavaria, Professor Steinheil was also contemporaneously engaged in demonstrating the practicability of another form of electric telegraph.

We have now arrived at a period from which the existence of this means of communication may be properly said to date; but the appliances hereafter to be described are more or less the offspring of these early labours.

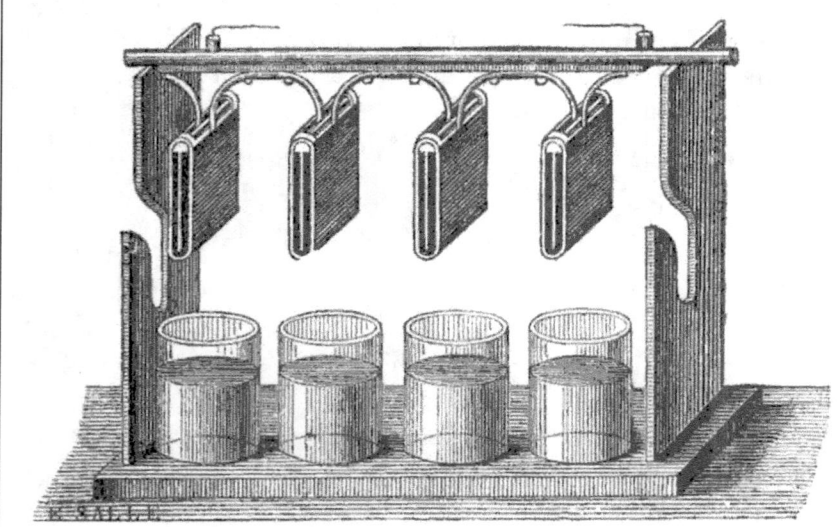

Wollaston's battery

4

THE PRODUCTION OF ELECTRICITY BY CHEMICAL DECOMPOSITION AND VARIOUS FORMS OF VOLTAIC BATTERIES ORDINARILY USED

69. Common plate battery—70. Combination of currents—71. Loss of power by imperfect conduction—72. Cylindrical batteries—73. Pairs of battery plates, or cells, and poles defined—74. Origin of term voltaic pile—75. Use of sand in charging batteries—76. To vary intensity of current 77. Description of batteries—Wollaston's, Hare's, Daniel's, Grove's, Smee's, Bunsen's, Marie Davy's.

69. ONE of the most simple forms of voltaic battery is that represented in Figure 18, which consists of a glazed earthenware trough, **A B**, divided by partitions into a series of parallel cells, and a series of zinc and copper plates, **A' B'**, of shape and magnitude corresponding with the cells, attached to a wooden support, each copper plate being connected at the top, under the wood, by a band of metal, with the zinc plate which immediately succeeds it in the series. For brevity, let us designate the first copper plate, C_1, the second, C_2, the third, C_3, and so on, proceeding from **A'** towards **B'**, and let the first zinc plate, which is connected with C_1, by a metal band, be called Z_2, the next, which is similarly connected with C_2, be called Z_3, and so on from **A'** towards **B'**. Now, the intervals between the plates being so arranged as to correspond with the width of the cells, the series of plates may be let down into the cells so that a partition shall

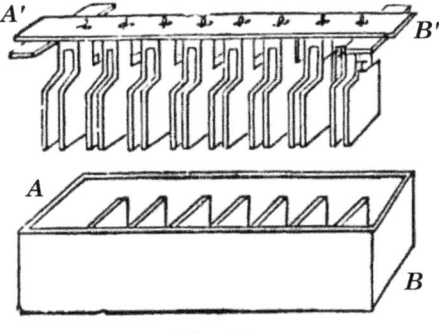

Fig. 18
Simple voltaic battery

separate every pair of plates which are connected by a metal band. Thus, the first partition will pass between C_1, and Z_2, the second between C_2 and Z_3, the third between C_3 and Z_4, and so on. It appears, therefore, that the first cell proceeding from A towards B will contain only the copper plate C_1, the second will contain C_2 and Z_3, the third, C_3 and Z_4, and so on, the last cell at the extremity B of the series containing only the last zinc plate, which we shall call Z_n.

Now, it is evident that as the arrangement thus stands, the first and last cells of the series would differ from the intermediate ones, inasmuch as, while each of the latter contains a pair of plates, each of the former contains only a single plate, the first copper C_1, and the last zinc Z_n. To complete the arrangement, therefore, it will be necessary to place a zinc plate, which we shall call Z_1, in the first cell to the left of C_2, and so as not to be in contact with it, and in like manner a copper plate, which we shall call C_n, in the last cell B to the right of Z_n, and so as not to be in contact with it. Let wires be soldered to the upper edges of these terminal plates Z_1 and C_n, and let them be carried to any desired distances, but finally connected with plates, or any other masses of metal, buried in the ground at n' and p', Figure 19.

These dispositions being made, let us suppose the cells to be filled with a weak acid solution, such as has been already described, but so that the liquid in one cell may not overflow into the next. A current of electricity will now be established along the wire passing as indicated by the arrows, from the last copper plate at P, to the

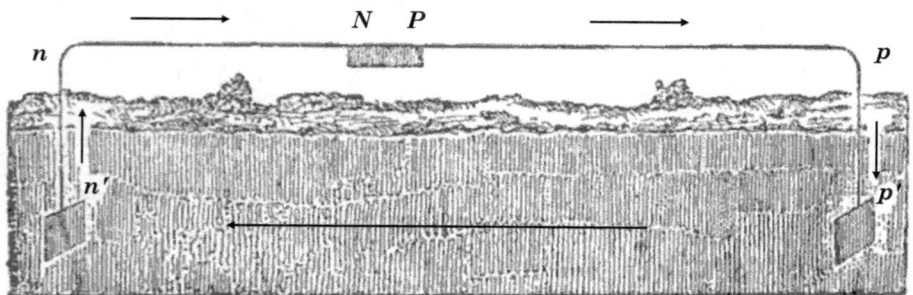

Fig. 19
Current flow of an earthed voltaic battery

earth at p', while a current will pass from the earth at n' to the first zinc plate Z_1, at N.

This current is produced by the combined voltaic action of all the pairs of plates contained in the cells of the trough.

70. The current produced by the combination Z_1, C_1, in the first cell, will flow from the plate C_1, by the band of metal to the plate Z_2, in the second cell. It will follow this course because of the conducting power of the metals, and the insulating power of the wood and earthenware, which prevents its escape. From the plate Z_2, it will pass through the acidulated water to the plate C_2, for although this water has not a conducting power equal to that of metal, it has nevertheless sufficient to continue the current to C_2. From C_2 it will pass by the band of metal to Z_3, and from that through the liquid in the third cell to C_3, and from that by the metal to Z_4, and so on until it arrives at the last plate C_n, of the series, from which it will pass, by the conducting wire, from P to p'.

It is evident, therefore, that the current produced by the voltaic combination in the first cell must pass successively through the plates and liquid in all the cells before it can arrive at P.

In the same manner it may be shown that the current produced in the second cell containing Z_2 and C_2, must pass through all the succeeding cells before it can reach P, and so of all the others.

71. Now, if the metals and liquid were perfect conductors, each of these currents would arrive at P with undiminished force, and then the current upon the wire $P\ p'$ would be as many times more intense than a current produced by a single voltaic combination as there are cells. But this is not so. The metals copper and zinc, though good conductors, are not perfect ones, and the acidulated water is a very imperfect one. The consequence is, that the currents severally produced in each of the cells, suffer a considerable loss of force before they arrive at the conducting wire $P\ p'$; and mathematical formulae, based on theoretical principles and practical data, have been contrived to express in each case the effects of this diminution of force due to the imperfect conducting power, or the resistance, as it has been called, of the elements of the battery.

Without going into the reasoning upon which these investigations are founded, it will be sufficient for our present object to state, that in all cases, a current of greater or less force is transmitted to the terminal plate of the series from each of the cells, no matter how numerous they may be, and in some cases batteries have been constructed and brought into operation, in scientific researches, which consisted of as many as two thousand pairs of plates.

72. To simplify the explanation, as well as because the form described is very generally used for telegraphic purposes, we have here selected the plate battery to illustrate the general principle upon which all voltaic combinations are founded. In Figure 20 is represented the disposition of the cylinders in a battery formed on

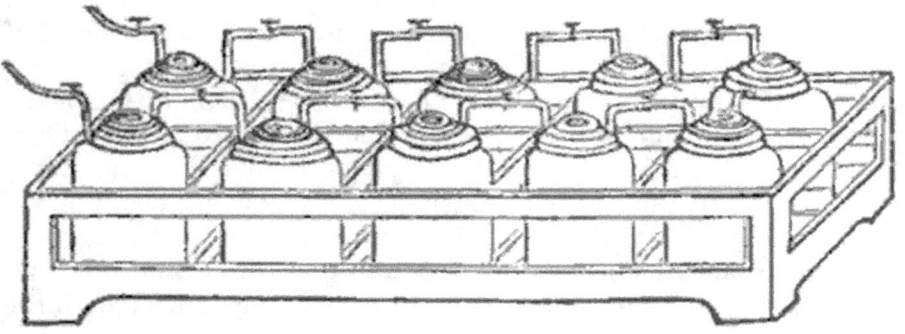

Fig. 20
General arrangement of a Daniel or Grove battery

the principles of Daniel or Grove, where the metallic connection of each copper or charcoal element of one pair, with the zinc element of the succeeding pair, is represented by a rectangular metallic bar or wire.

73. Each combination of two metals, or of one metal and charcoal, which enters into the composition of a battery, is usually called a PAIR, or an ELEMENT, and, in practical telegraphy, a *cell*. Thus, a battery is said to consist of so many PAIRS, or ELEMENTS, or cells.

The end of the battery from which the current issues is called its POSITIVE POLE, and that to which it returns is called its NEGATIVE POLE. Thus, in the batteries explained above, *P* is the positive, and *N* the negative pole.

Since in the most usual elements, zinc and copper, the current issues from the last copper plate, and returns to the first zinc plate, the positive pole is sometimes called the COPPER POLE, and the negative the ZINC POLE.

74. The voltaic battery is sometimes called the VOLTAIC PILE. This term had its origin from the forms given to the first voltaic combination by its illustrious inventor. The first pile constructed by Volta was formed as follows:— A disc of zinc was laid upon a plate of glass. Upon it was laid an equal disc of cloth or pasteboard, soaked in acidulated water. Upon this was laid an equal disc of copper. Upon the copper were laid, in the same order, three discs of zinc, wet cloth, and copper, and the same superposition of the same combinations of zinc, cloth, and copper, was continued until the pile was completed. The highest disc (of copper) was then the positive, and the lowest disc (of zinc) the negative pole, according to the principles already explained.

It was usual to keep the discs in their places by confining them between rods of glass. Such a pile, with conducting wires connected with its poles is represented in Figure 21.

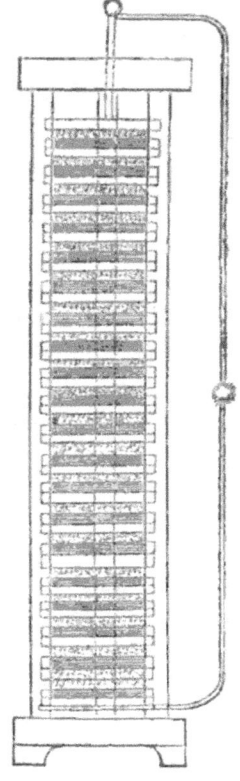

Fig. 21
A voltaic pile

75. As the batteries used on telegraphic lines are liable to frequent removal from place to place while charged with the acidulated water, or other exciting liquid, it has been found desirable to contrive means to prevent such liquid from being spilled, or thrown from cell to cell. This has been perfectly accomplished by the simple expedient of filling the cells with silicious sand, which is kept saturated with the exciting liquid so long as the battery is in operation.

76. It is often necessary, in telegraphic operations, to vary the intensity of the current. This is accomplished, within certain limits, without changing the battery, in the following manner:—

If it be desired to give the full force of the battery to the current, the wires are attached to the terminal plates, so that the entire battery is between them. But if any less intensity is desired, the wires, or one of them, is attached to intermediate plates, so that they shall include between them a part only of the battery. The part included between them is alone active in producing the current, all the elements which are outside the wires being passive. The battery, in effect, is converted into one of fewer elements.

Provisions are made, which will be explained hereafter, by which the operator can, by a touch of the hand, thus vary the force of the battery.

77. A very simple and inexpensive form of battery, suggested by the late Dr. Wollaston, is generally used in England, particularly in connection with the Railway telegraphs. It consists of plates of zinc and copper joined together by a band of copper, placed in a long wooden trough divided into cells by transverse partitions. The cells are usually filled with sand, moistened with water mixed with sulphuric acid in the proportion of about one part of strong acid to fifteen of water. A more intense current could be produced by using a stronger solution, but it is found preferable to augment its intensity by increasing the number of plates in the battery. The dimensions of the plates are generally four to five inches wide, and three to four inches deep. The thickness of the zinc plates is something less than a quarter of an inch. The cells are filled with sand to within an inch of the top, and the parts of the plates above the sand are varnished as a protection against corrosion, and to keep them clean. In general, the troughs are made either of glazed gutta percha or some compact wood, such as oak, or teak, made water-tight by cement or marine glue. When the trough is wood the partitions of the cells are slate, the width of each cell being one inch and a quarter to one inch and a half. The troughs contain, some twenty-four, and some twelve cells.

Glass troughs, with alternate porous divisions of earthenware, are found to answer very well with Daniel's form of battery, being both durable, and constant in action.

A very ingenious voltaic combination has been invented by Marie Davy, in which moistened bisulphide of mercury is used as a medium, zinc and graphite plates being employed. The mercury in the salt is gradually eliminated by electric action, keeping by electrolytic deposit the surface of the zinc plate amalgamated, and finally leaving a deposit of mercury in the cell. This battery is expensive at first, but the mercury deposited ultimately reduces the cost, and the wear of the zinc plates is very small. It has been introduced on some of the French and English lines. The graphite plates are cut from the refuse of gas retorts, and covered with a coating of powdered platinum by electro-chemical deposition. A sketch of this battery is shown in Figure 22. Batteries of this sort, consisting of twenty-four cells, give a current of sufficient force for an overground line of wire of 50 miles. For 100 miles, 48 cells, and for 200 miles, three troughs of 24 cells are required. These batteries may frequently give superfluous force, but it is necessary to provide against the contingency of leakage by accidental defects of insulation.

The durability of these batteries is increased by amalgamating the zinc plates. This is effected by first washing them in acidulated water, and then immersing them in a bath of mercury for one or two minutes. The mercury will combine with the zinc and form a superficial coating of the amalgam of zinc. When they are worn by use, they may be restored, by scouring them, and submitting them to the same process, and this may be continued until the zinc become too thin to hold together.

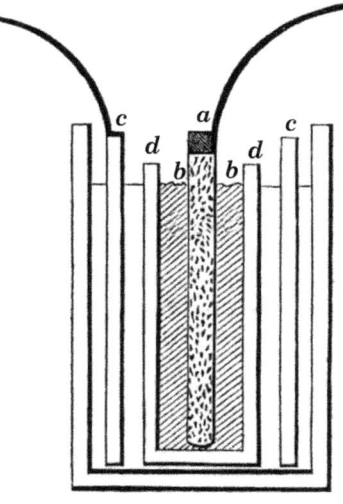

Fig. 22
Marie Davy battery

a   Graphite platinised.
b b Bisulphide of mercury
c   Circular zinc plate.
d   Porous cell.

New batteries, when carefully put together, will, with care, do duty for six or eight months, when the work is not heavy; but on a busy line they become exhausted in three or four months.

The late Dr. Wollaston proposed an arrangement, in which the copper plate was bent into two parallel plates, a space between them being left for the insertion of the zinc plate, the contact of the plates being prevented by the inter-position of bits of cork or other non-conductor. The system thus combined was immersed in dilute acid, contained in a porcelain vessel.

Dr. Hare of Philadelphia contrived a voltaic arrangement, consisting of two metallic plates, one of zinc and the other of copper, of equal length, rolled together in the form of a spiral, a space of a quarter of an inch being left between them. They are maintained parallel without touching, by means of a wooden cross at top and bottom, in which notches are provided at proper distances, into which the plates are inserted, the two crosses having a common axis. This combination is let into a glass or porcelain cylindrical vessel of corresponding magnitude, containing the exciting liquid.

This arrangement has the great advantage of providing a very considerable electromotive surface with a very small volume. The exciting liquid recommended for these batteries when great power is desired, is a solution in water of 2¼ per cent, of sulphuric, and 2 per cent, of nitric acid. A less intense but more durable action may be obtained by a solution of common salt, or of 3 to 5 per cent, of sulphuric acid only.

It is not essential that the water in which the metals are immersed be acidulated, as we have supposed, by sulphuric acid. Any acid which will promote the oxidation of the zinc without affecting the copper will answer. Nor is it indeed necessary that any acid whatever be used. A saline solution is often found more convenient. Thus common salt dissolved in the water will produce the desired effect.

Of the various voltaic combinations which have been applied in scientific researches, four only have been found available to any considerable extent in the working of electric telegraphs, the zinc and copper plate combination described above, Daniel's constant battery, Grove's battery, Bunsen's modification of it, and the magneto-electric apparatus.

Daniel's combination, which is extensively used in working telegraphs, consists of a copper cylindrical vessel *C C,* Figure 23, widening near the top *a d.* In this is placed a cylindrical vessel of unglazed porcelain *p.* In this latter is placed the hollow cylinder of zinc **Z**, already described. The space between the copper and porcelain vessels is filled with a saturated solution of the sulphate of copper, which is maintained in a state of saturation by crystals of the salt placed in the wide cup **a b c d**, in the bottom of which is a grating composed of wire carried in a zigzag direction between two concentric rings, as represented in plan at **G**. The vessel **p**, containing the zinc, is filled with a solution of sulphuric

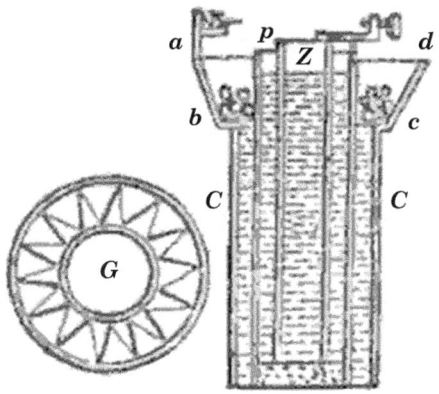

Fig. 23
Daniel's constant battery

acid, containing from 10 to 25 per cent. of acid when greater electromotive power is required, and from 1 to 4 per cent. when more moderate action is sufficient. This form of battery evolves electricity of low tension, but is found very economical in use from the long time it lasts without renewal, and is found especially suited for telegraphs from its cheapness and the invariable nature of its current.

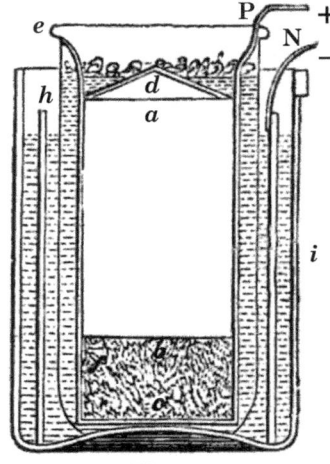

Fig. 24
Pouillet's modification of
Daniel's battery

The following modification of Daniel's system was adopted by M. Pouillet in his experimental researches, and is the form and arrangement used in France for the telegraphs. A hollow cylinder **a**, Figure 24, of thin copper, is ballasted with sand **b**, having a flat bottom **c**, and a conical top **d**. Above this cone the sides of the copper cylinders are continued, and terminate in a flange **e**. Between this flange and the base of the cone, and near the base, is a ring of holes. This copper vessel is placed in a bladder which fits it loosely like a glove, and is tied round the neck under the flange **e**. The saturated solution of the sulphate of copper is poured into the cup above the cone, and, flowing through the ring of holes, fills the space between the bladder and the copper vessel. It is maintained in its state of saturation by crystals of the salt deposited in the cup.

This copper vessel is then immersed in a vessel of glazed porcelain *i,* containing a solution of the sulphate of zinc or the chloride of sodium (common salt). A hollow cylinder of zinc *h*, split down the side so as to be capable of being enlarged, or contracted at pleasure, is immersed in this solution surrounding the bladder. The poles are indicated by the conductors **P** and **N**, the positive proceeding from the copper, and the negative from the zinc.

M. Pouillet states that the action of this apparatus is sustained without sensible variation for entire days, provided the cup above the cone *d* is kept supplied with the salt, so as to maintain the solution in the saturated state.

In the batteries used for the telegraphs on the French railways, the liquid in which the zinc cylinder is immersed is pure water, and this is found to answer in a very satisfactory manner.

The current flows from the copper cylinder and returns as usual to the zinc.

Grove's battery consists of two liquids, sulphuric and nitric acids, and two metals, zinc and platinum, arranged in the following manner:–

A hollow cylinder of zinc *Z Z*, Figure 25, open at both ends as already described, is placed in a vessel of glazed porcelain, *V V*. Within this is placed a cylindrical vessel *v v*, of unglazed porcelain, a little less in diameter than the zinc *Z Z*, so that a space of about a quarter of an inch may separate their surfaces. In this vessel *v v*, is inserted a cylinder *C C* of platinum, open at the ends, and a little less than *v v*, so that their surfaces may be about a quarter of an inch asunder. Dilute sulphuric acid is then poured into the vessel *V V*, and concentrated nitric acid into *v v*; **P** proceeding from the platinum will then be the positive, and **N** proceeding from the zinc the negative pole.

In Smee's battery, plates of silver instead of copper are used in combination with zinc, the exciting fluid is more strongly acidulated than is customary in the Wollaston battery, and the elec-

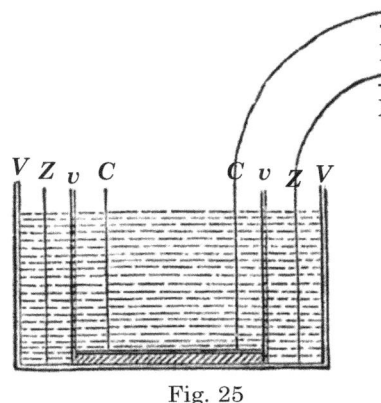

Fig. 25
Grove's battery

trical action is consequently more powerful. The Grove and Smee batteries are usually so arranged that the plates are lifted out of the cells by turning the handle of a rack to which they are fixed, so as to prevent the decomposition of the plates when the batteries are not actually in use. These two forms of voltaic arrangement are, however, expensive in construction and working.

Bunsen contrived a battery which has taken his name, and which, while it retains all the efficiency of Grove's, can be constructed at much less expense, the platinum element being replaced by the cheaper material of charcoal.

In the vessel v v is inserted, instead of a hollow cylinder of platinum, a solid cylindrical rod of charcoal, made from the residuum taken from the retorts of gasworks. A strong porous mass is produced by repeatedly baking the pulverised coke, to which the required form is easily imparted. Dilute sulphuric acid is then poured into the vessel V V, and concentrated nitric acid into v v. The electric fluid issues from a wire connected with the charcoal, and returns by one connected with the zinc.

The chief advantage of Daniel's system is that from which it takes its name, its *constancy*. Its power, however, in its most efficient state, is greatly inferior to that of the carbon or platinum systems of Bunsen and Grove. A serious practical inconvenience, however, attends all batteries in which concentrated nitric acid is used, owing to the diffusion of nitrous vapour, and the injury to which the parties working them are exposed by respiring it. In my own experiments with Bunsen's batteries the assistants have been often severely affected; and in addition to this, the production of great power is necessarily accompanied by a speedy consumption of the materials of the battery, and consequently an increased expense.

In the use of the platinum battery of Grove, the nuisance produced by the evolution of nitrous vapour is sometimes mitigated by enclosing the cells in a box, from the lid of which a tube proceeds which conducts these vapours out of the room.

Although each of the simple combinations described above would produce an electric current, which, being transmitted upon a conducting wire, would be attended with effects sufficiently distinct to manifest its presence, such a current would be too feeble in its intensity to serve the purposes of a telegraphic line; and as no other simple voltaic combination yet discovered would give to a current the necessary intensity, the object has been attained by placing in connection a series of such combinations, in such a manner that the currents produced by each of them being transmitted in the same direction, on the same conducting wire, a current having an intensity due to such combination may be obtained.

To each of these forms of battery may be assigned its special use — The Daniel's, Wollaston's, and Davy's for telegraphs; Bunsen's for electro-plating, or where a large volume of electricity of low tension is required; and Grove's and Smee's in experimental researches when the most speedy and energetic electric action is needed, such as in the production of the electric light, deflagration of metals, ignition of gunpowder for mining purposes, &c.

PLATE V

Interior of message room in the Electric Telegraph Office,
Charing Cross, London

5

TELEGRAPH CONDUCTING WIRES AND THE VARIOUS MODES OF INSULATING THEM

78. HAVING explained, generally, the manner in which the electric current is produced and maintained, I shall now proceed to explain the various expedients by which it is conducted from station to station, along the telegraphic line, and by which injurious waste by leakage or drainage is prevented or diminished.

The conducting wires used for telegraphic lines are of iron, usually the sixth of an inch in diameter. On all European lines they are submitted to a process called galvanisation, being passed through a bath of liquid zinc, by which they become coated with that metal. This zinc surface being easily oxydizable, is soon, by the action of air and moisture, converted into the oxide of zinc, which, being insoluble by water, remains upon the wire, and protects the iron from all corrosion.

When a great length of wire is to be stretched between two distant points without intermediate support, steel wire is often preferred to iron, in consequence of its greater strength and tenacity.

Copper being a bettor conductor of electricity than iron, as well as being less susceptible of oxidization, would on these accounts be more eligible for telegraphic purposes. Its higher price, and the possibility of compensation for the inferior conducting power of iron, by using larger wire or greater battery power, has rendered it preferable to use that metal. Besides this, the copper wire, if suspended in any considerable length, stretches, while the iron does not, from its far greater rigidity.

79. When galvanised iron wires pass through large towns where great quantities of coal are burnt, the sulphurous acid gas resulting from such combustion acting upon the oxide of zinc which coats the conducting wire, converts it into a sulphate of zinc, which being soluble in water, is immediately dissolved by rain, leaving the iron unprotected. The wire consequently soon rusts, and is corroded. In some cases the telegraph wires are reduced by this cause to the thinness of a common sewing needle in a few years. As a partial protection against corrosion, the wires in towns are frequently painted.

The wires used on the American lines are of iron, similar to the European, but are not galvanised. They soon become coated with their own oxide. A pair of galvanised wires have been placed between New York and Boston, and I have been informed by Mr. Shaffner, the secretary of the American Telegraph Confederation, that at certain times during the winter, it has been found that they were unable to work the telegraph with these wires, while its operation with the wires not galvanised, was uninterrupted. Mr. Shaffner also states that several anomalous circumstances have been manifested upon some extensive lines of wire erected on the vast prairies of Missouri. Thus, in the months of July and August, it is found that the telegraph cannot be worked from two to six in the afternoon, being the hottest hours of the day. These circumstances are ascribed to some unexplained atmospheric effects.

The manner in which the conducting wires are carried from station to station is well known. Every railway traveller is familiar with the lines of wire extending along the side of the railways, which, when numerous, have been not inaptly compared to the series of lines on which the notes of music are written, and which are the metallic wires on which invisible messages are flying continually with a speed that surpasses imagination. These are suspended on posts, erected at intervals of about sixty yards, being at the rate of thirty to a mile. They therefore supply incidentally a convenient means by which a passenger can ascertain the speed of the train in which he travels. If he count the number of telegraph posts which pass his eye in two minutes, that number will express in miles per hour the speed of the train.

80. Since the current of electricity which flows along the wire has always a tendency to pass by the shortest route possible to the ground, it is evident that the supports of the wires upon these posts ought to possess, in the highest attainable degree, the property of insulation; for even though the entire stream of electrical fluid might not make its escape at any one support, yet if a little escaped at one and a little at another, the current would, in a long line, be soon so drained that what would remain would be insufficient to produce those effects on which the efficiency of the telegraph depends. Great precautions have therefore been taken, and much scientific ingenuity has been expended in contriving supports which shall possess, in the highest attainable degree, the property of insulation.

81. To each of these posts or poles are attached as many inverted cups, or other forms of support, in porcelain or glass, as there are wires to be supported. Each wire passes through a tube, or is supported on an inverted cup; the material of which being among the most perfect of the class of non-conducting substances, the escape of the electricity at the point of contact is impeded.

Notwithstanding various precautions of this kind, a considerable escape of electricity still takes place in wet weather. The coat of moisture which collects on the wire, its support, and the post, being a conductor, carries away more or less of the fluid. Consequently, more powerful batteries are necessary to give effect to the telegraph in wet than in dry weather. In England, and on the Continent, the material hitherto used for the support of the wires is principally a sort of earthen or stoneware. In the United States it is generally glass.

The forms of these insulating supports are various. The

penthouse or roof-shape has been usually preferred as preserving a dry channel between the wire and the post. The material used most commonly in England, a sort of brown stoneware, has the advantage, besides being a good insulator, of throwing off wet, as water falls from a duck's wing, leaving the surface dry. A pitcher of this ware, plunged in water, scarcely retains any moisture upon it.

82. The posts vary generally from 15 to 30 feet in height, the lowest wire being about ten feet above the ground, except in cases where greater height is required to allow vehicles to pass under it, as when the wires cross a common road, or pass from one side of the railway to the other. The poles are about 6 inches square at the top, and increase to 8 inches at the bottom. In some cases they are impregnated with certain chemical solutions, to preserve them from rotting, and are generally painted, the parts which are in the ground being charred and tarred. The manner of treatment, however, varies in different countries.

83. In Figures 26 and 27 are represented different forms of supports used in England. To cross-pieces of wood, **A A'**, bolted upon the post (Figure 26), are attached balls, **b**, of stoneware, as described above, in which grooves or slits are formed to receive and support the wires. These supports are protected from rain and from the deposition of dew by hoods of zinc-coated iron placed over them. Glass being so much better an insulator, balls of that material are recently being substituted for the stone ware. This mode of insulation was devised by Mr. Clark.

Another form of support,

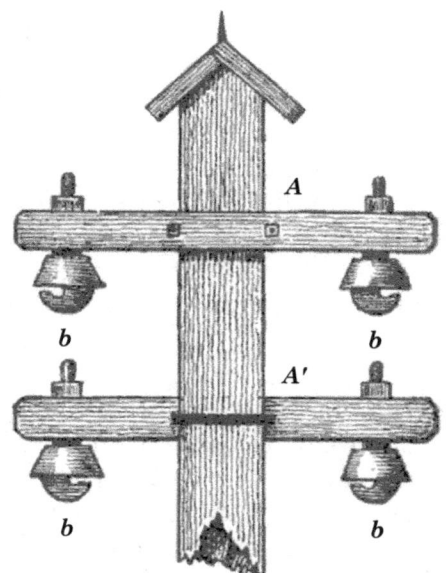

Fig. 26
Telegraph pole, cross arms and
stoneware insulators

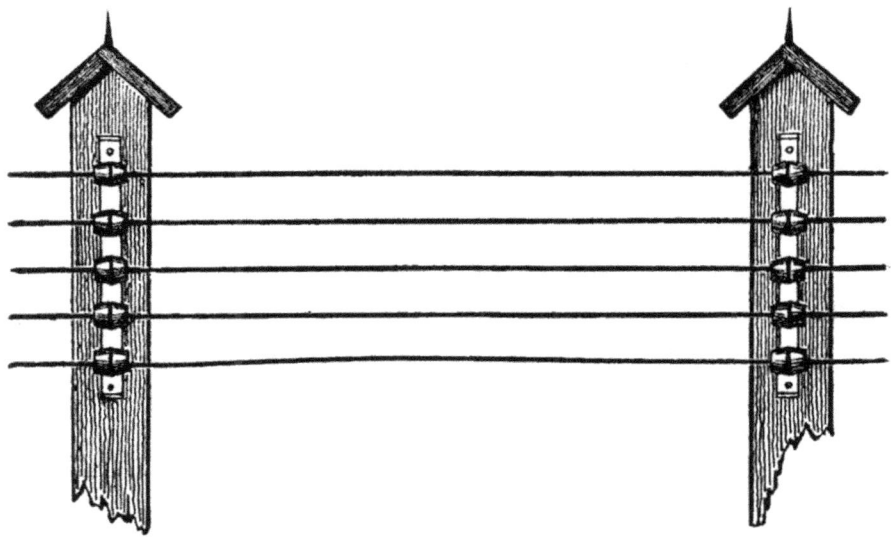

Fig. 27

Cooke and Wheatstone type poles and stoneware ring insulators

sheltered by a sort of sloping roof, is represented in Figure 27. On the front of the post is a wooden arm to which a series of stoneware rings are attached, through which the wires pass. These rings have the form of two truncated cones placed with their larger bases in contact. This form was the one first adopted by Messrs. Cooke and Wheatstone, but has now been given up.

It is usual, where the wires are numerous, as on some of the lines near London, to attach these supports both to the front and back of the post. Sometimes as many as thirty, or even more, insulators, each bearing a wire, are attached to a single telegraph post.

Apparatus for tightening the wires, as shown at Figure 28, were formerly attached to the posts at every half-mile, but are not now resorted to, the wire being bound to each insulator so that it cannot run out.

An improved system of insulation invented by Sir Charles Bright, is shown at Figure 29. The insulators are composed of glass or porcelain, somewhat in the form of an umbrella, with a slot at the top to hold the wire, which is fastened to each insulator by a turn

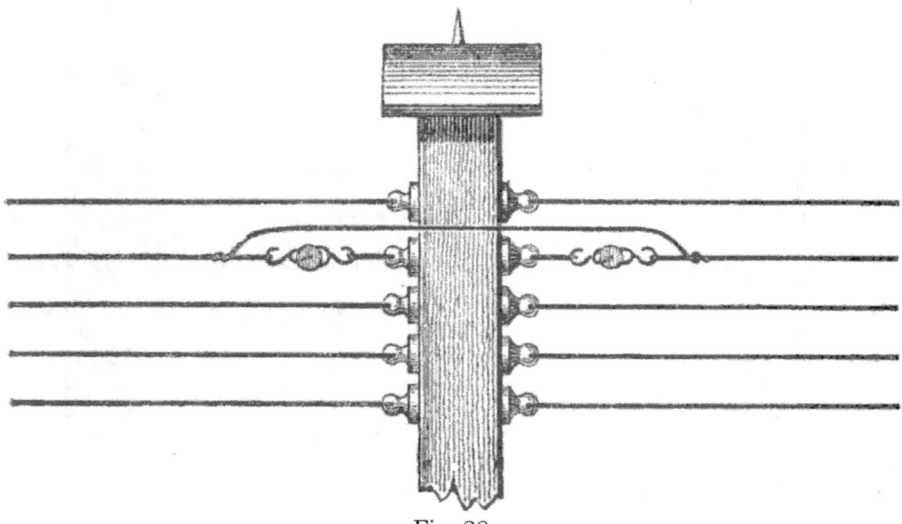

Fig. 28
Apparatus attached to the post for tightening the wires

of thin wire passed round its head. The arms on the post are increased in length upwards; so that in case an upper wire becomes accidentally broken and falls, it clears the wires below, and does not impede working by coming into contact with them.

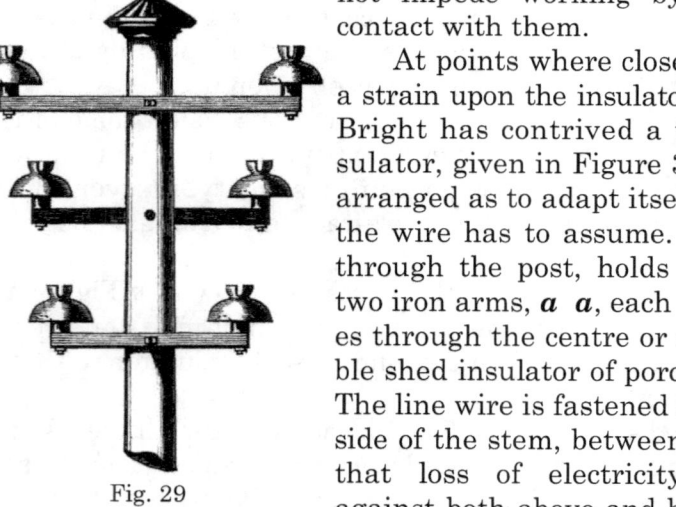

Fig. 29
Bright's improved system
of insulation

At points where close curves throw a strain upon the insulators, Sir Charles Bright has contrived a traversing insulator, given in Figure 30, which is so arranged as to adapt itself to any angle the wire has to assume. A bolt passes through the post, holds upon a hinge two iron arms, *a a*, each of which passes through the centre or stem of a double shed insulator of porcelain or glass. The line wire is fastened round the outside of the stem, between the sheds, so that loss of electricity is guarded against both above and below the wire. The hinge allows the insulator to trav-

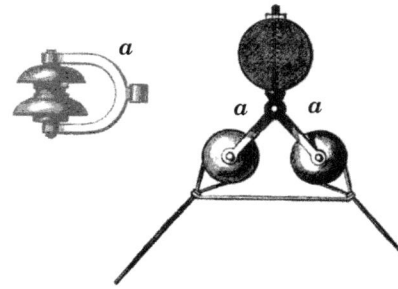

Fig. 30
Bright's traversing insulator

erse in any direction required, and the bolt is so fixed in the post that its line of direction bisects the angle of strain. This form of insulator has been generally adopted in this country and abroad, wherever long stretches of wire are required, as in crossing the streets of towns, &c.

Figure 31 shows a porcelain insulator which has been substituted in many places for the insulation described in Figures 26 and 27. In this case the iron bolt of the insulator fits into the socket of an iron bracket screwed to the post. The slot in the top of the porcelain cup is so curved as to give the wire a slight bend, and check it from running out in case of breakage.

Another still more recent plan is given in Figure 32, where two cups of earthenware are fitted one within the other, so that in case of leakage through one the other may be sound.

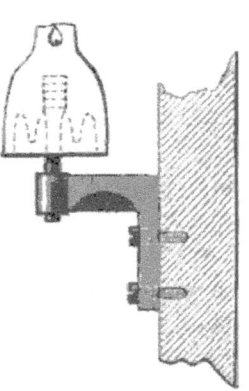

Fig. 31
Porcelain insulator

It was found that, in manufacturing earthenware or porcelain by the ordinary process of turning or moulding, the material was always more or less porous when the glazing deteriorated; and to obviate this Mr. Jobson, of Dudley, introduced a method of compressing the porcelain in moulds by a powerful hydraulic machine, which has proved a great advantage. It is customary to test each insulator during the process of manufacture, by soaking the cups in acidulated water for some days, and then applying a strong electrical power to ascertain if there is the slightest leakage apparent. The iron bolts of the insulators are usually galvanised, to prevent rusting, and are sometimes covered with vulcanised india-rubber or other

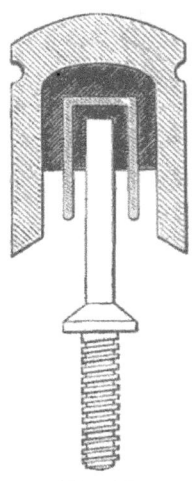

Fig. 32
Double cup
insulator

insulating material, to lessen the chance of leakage between the wire and bolt. Vulcanised india-rubber, or ebonite, has also been moulded into the cup or invert form, and used as an insulator for overground wires, but is found to become partially porous after being up a year or two.

84. In France the posts are from twenty to thirty feet high, placed at distances varying from sixty to seventy yards asunder, and sunk to a depth of from three to seven feet in the ground. They are impregnated with sulphate of copper to preserve them from rotting by damp.

The conducting wire rests in an iron hook, which is fastened by sulphur into the highest part of the cavity of an inverted bell, formed of porcelain, from which

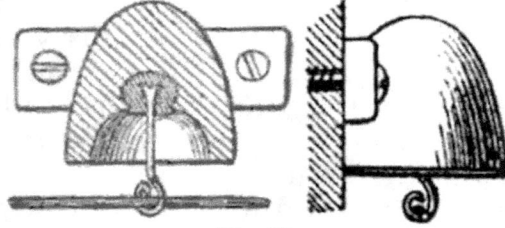

Fig. 33
French inverted bell insulator

two ears project, which are screwed to the post. A section and a side view of this apparatus is given in Figure 33.

The winding posts are placed at distances of a kilometre (six-tenths of a mile). The apparatus used for tightening the wire consists of two drums or rollers, each carrying on its axis a ratchet wheel with a catch. These drums are mounted on iron forks formed at the ends of an iron bar, which is passed through an opening in a porcelain support, and secured in its position by pins, the porcelain support being attached to the post by screws passing through ears projecting from it.

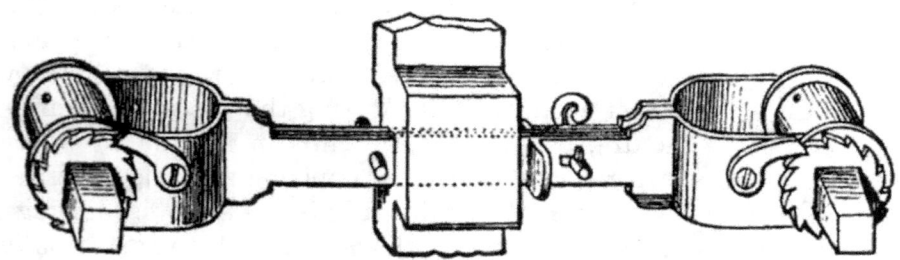

Fig. 34
French wire tightening apparatus

A front view of this winding apparatus is given, Figure 34; a side view of the porcelain support, showing the opening through which the iron bar is passed, and the screws by which it is attached to the post, is given in Figure 35.

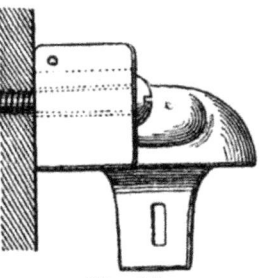

Fig. 35
Porcelain support

The conducting wires used in France are similar to those used on the English lines.

85. The insulating supports of the wires used on the American lines are very various in form.

The supports upon the principal Morse lines consist of a glass knob, *A*, Figure 36, upon which two projecting rings are raised in the groove between which the wire is wrapped. This glass knob is attached to an iron shank *B*, which is driven into the post.

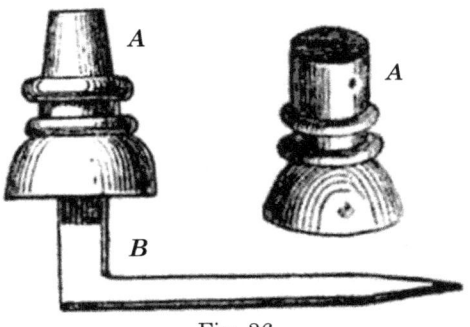

Fig. 36
American glass insulator and iron shank

Another form of support used on these lines is represented in Figure 37, which consists of two rectangular blocks of glass, in each of which is a semi-cylindrical groove corresponding with the thickness of the conducting wire, so that the wire being laid in the groove of one of them and the other being laid upon it, will be completely enclosed within the block of glass produced by their union. These blocks of glass are surrounded and protected by a larger block of wood, as represented in the figure, where the white part represents the glass, and the shaded part the wood.

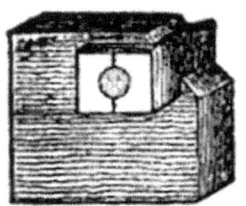

Fig. 37
Wood block and glass insulator

Such a form of insulation as this would be quite insufficient upon circuits exceeding 20 or 30 miles in length, and was probably an expedient adopted in the early days of telegraphy, before the principles of such work were understood.

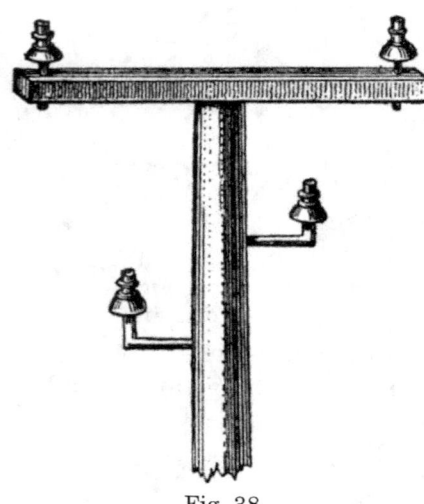

Fig. 38
American type wire support

The supports are sometimes attached to the sides of the posts, and sometimes placed upon a horizontal cross bar, as represented in Figure 38.

The supports used in House's lines consist of a glass cap about five inches length and four inches in diameter, having a coarse screw-like surface cut inside and out. This glass cap *A* Figure 39, is screwed and cemented into a bell-shaped iron cap *B* from three to four pounds in weight, projecting an inch below the lower edge of the glass, protecting it from being broken; this is fitted with much care to the top of the pole *C*, and is covered with paint or varnish. The conducting wire is fastened to the top of the cap by projecting iron points, and the whole of the iron cap is thus in the circuit, as the wire is of iron, and not insulated. To prevent the deposit of moisture, the glass is covered by a varnish of gum-lac dissolved in alcohol, and the ring-like form of the glass is to cause any moisture to be carried to the edge, and there drop off.

86. One of the forms of insulating support used on the German lines is represented in Figure 40, and consists of an insulating cap placed on the tapering end of a post *T*. The post terminates in a point *c*, an inch and a half in length and about six lines in diameter; this pole is covered with a porcelain cap *d d*, a sort of reversed cup; on its summit *e* there is a hole *a* inlaid with lead, in which the conducting wire *b b* enters; this insulator is then covered with a roof.

In localities where the insulators are exposed to the chance of

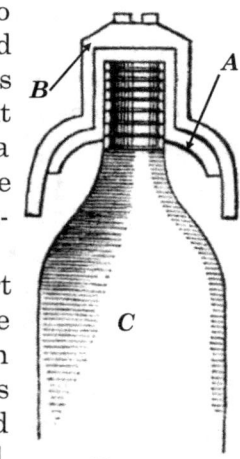

Fig. 39
Glass screwed insulator with iron cap

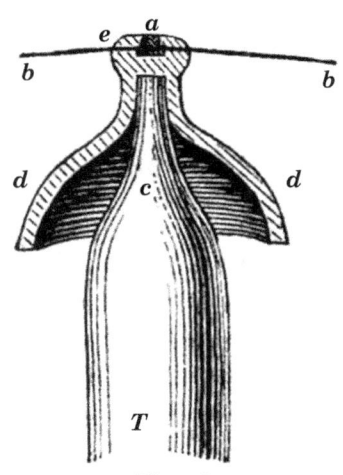

Fig. 40
German insulating support

injury, either from stone-throwing (a favourite amusement of lads in many districts), or other causes, an iron or teak cover is fitted on over the insulating cup of porcelain or glass.

It may be asked what prevents the escape of the electric fluid from the surface of the wire between post and post? In general, when wires are used on a smaller scale for the transmission of electric currents, the escape of the fluid is prevented by wrapping them with silk or cotton thread, which thus forms a non-conducting cover upon them, but on the scale on which they are used on telegraphic lines the expense of this, independently of the difficulty of protecting such covering from destruction by weather, would render it inadmissible.

The atmosphere, when dry, is a good non-conductor; but this quality is impaired when it is moist. In ordinary weather, however, the air being a sufficiently good non-conductor, a metallic wire will, without any other insulating envelope except the air itself, conduct the stream of electricity to the necessary distances. It is true that a coated wire, such as we have described, would be subject to less waste of the electric fluid en route; but it is more economical to provide batteries sufficiently powerful to bear this waste, than to cover such extensive lengths of wire with any envelope.

87. Atmospheric electricity having been found to be occasionally attracted to the wires, and to pass along them, so as to disturb the indications of the telegraphic instruments, and sometimes even to be attended with no inconsiderable danger to those employed in working the apparatus; various expedients have been contrived for removing the inconvenience and averting the danger. The current produced by this atmospheric electricity is often so intense as to render some of the finer wires used in certain parts of the apparatus at the stations, red hot, and sometimes even to fuse them. It also produces very injurious effects by demagnetising the needles, or

imparting permanent magnetism to certain bars of iron included in the apparatus, which thus become unfit to use.

One of the expedients used for the prevention of these inconvenient and injurious effects is to place common lightning conductors on the posts. The points of these are shown upon the posts in Figures. 26 and 27.

Mr. Walker, of the South-Eastern Company, and M. Breguet, of Paris, have each invented an instrument for the better protection of telegraphic stations from atmospheric electric discharges. Both these contrivances have been found in practice to be efficacious, and though differing altogether in form they are similar in principle. In both, a much finer wire than any which lies in the regular route of the current is interposed between the line wire and the station, so that an intense and dangerous atmospheric current must first pass this fine wire before reaching the station. Now it is the property of such a current to raise the temperature of the conductor over which it passes to a higher and higher point in proportion to the resistance which such conductor offers to its passage. But the resistance offered by the wire is greater in the same proportion as its section is smaller. The safety wire interposed in these contrivances is, therefore, of such thinness that it must be fused by a current of dangerous intensity. The wire being thus destroyed, all electric communication with the station is cut off, and the extent of the inconvenience is the temporary suspension of the business of the line until the breach has been repaired.

To obviate the inconvenience arising from the interruption of communication by the fusing of the thin wire in the ordinary form of lightning protectors, an apparatus was invented some time since by Sir Charles Bright and the writer, in which the fact that any intense form of electricity, such as lightning, will pass with ease through rarefied air, is turned to account. This instrument is shown in Figure 41, where *a* is a small air-tight chamber of glass or other suitable material; *b*, a metal terminal plate connected with the line wire outside the telegraph instrument; and *c*, a metal plate connected to the earth. The air within the box is sufficiently exhausted, by means of an air-pump, to allow lightning to pass freely, and thus, when joined to a line of telegraph, this ap-

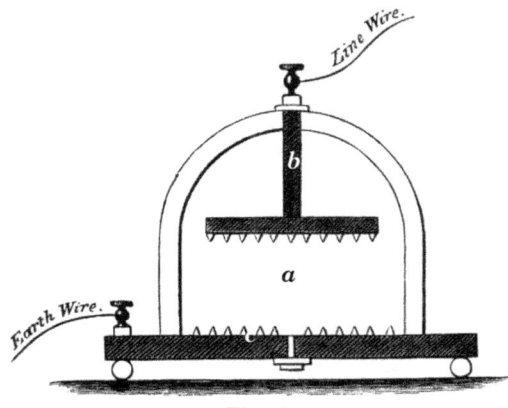

Fig. 41
Bright's rarefied air lightening protector

a. Glass chamber, from which the air has been partially exhausted
b. Metallic discharger from line wire
c. Metallic base of protector, connected with the earth

paratus acts as a safety-valve; not interfering with the ordinary electricity used in working the wires, but affording an easy channel to earth for lightning.

Expedients are used on the American lines to divert the atmospheric electricity from the wires, consisting merely of a number of fine points projecting from a piece of metal connected with the earth by a rod of metal. These points are presented to a metal plate, or other surface, attached to the line wire at the place where it enters the station. It is found that these points attract the atmospheric electricity, which passes to the ground by the conductor connected with them, but do not attract the electricity of the battery current.

88. The wires extended from post to post are continued in passing the successive stations of the line. The expedients by which the current is turned aside from the main wire, and made to pass through the telegraphic office of the station, differ more or less in their details on different lines and in different countries, but are founded on the same general principles. It will therefore be sufficient here to describe one of those commonly used on the British lines.

The conducting wire of the main line in passing the station is cut and the ends jointed by a shackle, as represented in Figure 28, in the case of a winding-post. This shackle breaking the metallic continuity would stop the course of the current. A wire is attached to the line wire below the shackle so as to receive the current which the latter would stop, and is carried on insulating supports into the telegraphic office, and put in connection with the telegraphic instrument. Another wire connected with the other side of the

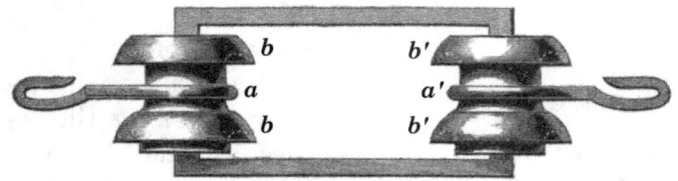

Fig. 42
Bright's shackle

instrument receives the current on leaving it, and being carried back on insulating supports to the line wire, is attached to the latter above the shackle, and so brings back the current, which continues its progress along the line wire.

A very effective form of shackle, the invention of Sir C. Bright, has been used for many years on the English lines. It is shown in Figure 42. The line wire is attached on each side to the hooks *a a'*, which are insulated from the iron frame-work and from each other by the double porcelain cups *b b'*. The connecting wires are then led into the station from each end of the line wire.

89. Although the mode of carrying the conducting wires at a certain elevation on supports above the ground has been the most general mode of construction adopted on telegraphic lines, it has been found in certain localities subject to difficulties and inconvenience, and some projectors have considered that in all cases it would be more advisable to carry the conducting wires underground.

This underground system has been adopted in the streets of London and of some other large towns.

The wires were at first wrapped with cotton thread, and coated with a mixture of tar, resin, and grease; but this was found a very defective expedient, and upon the introduction of gutta-percha, resin and its concomitants were at once dispensed with.

The wires are carried in iron pipes under the foot pavements, along the sides of the streets, and are thus conducted to the terminal stations of the various railways, where they are united to the lines of wire supported on posts along the sides of the railways, already described.

Provisions, called *testing posts*, are made at intervals of a quarter of a mile along the streets, by which any failure or accidental irregularity in the buried wires can be ascertained, and the place of such defect always known within a quarter of a mile.

Some of the wires of the British and Irish Magnetic Telegraph Company were at first laid and protected in the following manner.

Ten conducting wires are enveloped in a covering of gutta-percha, so as to be completely separated one from another. Thus prepared, they are deposited in a square creosoted wooden trough, measuring three inches in the side, so that nearly a square inch of its cross section is allowed for each of the wires. This trough is deposited on the bottom of a trench cut two feet deep along the side of the common coach road. A galvanised iron lid, of about an eighth of an inch thick, is then fastened on by clamps or small tenter-hooks, and the trench filled in. A section of the trough is given in Figure 43

The method of laying the wires in the streets adopted by this company is a little different. In this case iron pipes are laid, but they are split longitudinally. The under halves are laid down in the trench, and the gutta-percha covered wires being deposited, the upper halves of the pipes are laid on and secured in their places by means of screws through flanges left outside for the purpose.

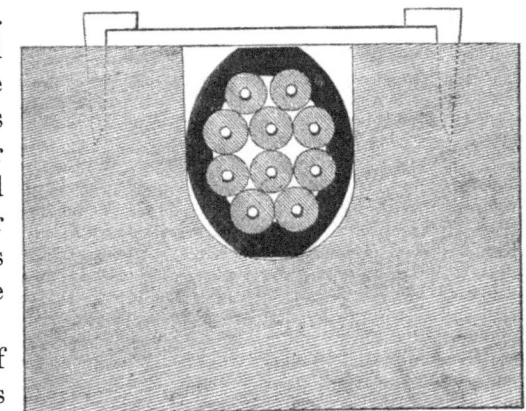

Fig. 43
Creosoted deal troughing with galvanised iron lid

To deposit the rope of gutta-percha covered wires in the trough, it is first coiled upon a large drum, which being rolled along slowly and uniformly over the trench, the rope of wires is payed off easily and evenly into its bed.

So well has this method of laying the wires succeeded, that in Liverpool the entire distance along the streets, from Tithebarn Railway Station to the Telegraph Company's offices in Exchange Street East, was laid in eleven hours; and in Manchester the line of streets, from the Salford Railway Station to Ducie Street, Exchange, was laid in twenty-two hours. This was the entire time occupied in opening the trenches, laying down the telegraph wires, refilling the trenches, and relaying the pavement.

90. In passing through tunnels the overground wires were subject to great inconvenience, owing to the quantity of water percolating through the roof, constantly failing on the wires and their supports, and thus injuring their insulation. It has been found that from this cause the current transmitted along one wire has been subject to leakages, a part of it passing by the moisture which surrounds the supports to an adjacent wire, so that being thus divided, part either returns to the station from which it has been transmitted, or goes on to a station for which it is not intended.

This inconvenience has been removed by adopting for tunnels the underground system. Mr. Walker, to whom great experience in the practical business of electric telegraphy, and considerable scientific knowledge, must give much authority on such a subject, at first adopted (apparently with very favourable results) a method of covering the wires, which pass through tunnels, with a coating of gutta-percha. The conducting wire thus treated was copper wire No. 16. The gum being well cleaned and macerated by steam, is put upon the wire by means of grooved rollers. Since this plan was first tried great improvements have taken place in the manufacture of gutta-percha covered wire; and under the superintendence for many years of the late Mr. Statham, and subsequently of Mr. Chatterton, vast quantities of wire so insulated have been provided for submarine cables, under-ground lines, and tunnels. The gutta-percha, after being thoroughly macerated and cleansed by several processes, is placed in a warm and plastic condition within a cylinder. The wire to be covered is drawn through a small circular opening, or die, in front of the cylinder. The pressure of a piston upon the gutta-percha in the cylinder forces it out with the wire in the centre. After being cooled in a tank of water, a second coating of gutta-percha is applied in a similar manner; and for important submarine lines as many as five coatings have been added to one another, to prevent the possibility of leakage of electricity through one or more of them.

Between 1851 and 1854 considerable lengths of gutta-percha covered wire were laid underground, both in England and Prussia, along the roads on some of the principal routes of the telegraph lines. The gutta-percha was, however, found to decay after a few years, and this plan of telegraph has been given up for long lines, overground wires being substituted.

91. Contact with air desiccates the gum, after a time, by absorbing its essential oil; the gutta-percha thus becomes to a certain extent porous and spongy, and no longer affords a sufficient insulation for the electricity. In some cases microscopic fungi are developed on the decaying material.

Water, however, and contact with vegetable tar, are found to act as preservatives; and in submarine cables, laid even fifteen years ago, the gutta-percha surrounding the wires is found to be as fresh and sound as when the cables were first manufactured.

92. In France and in the United States the wires, even in the cities and towns, are conducted on rollers at an elevation, as on other parts of the lines. In Paris, for example, the telegraphic wires proceeding from the several railway stations are carried round the external boulevards and along the quays, the rollers being attached either to posts or to the walls of houses or buildings, and are thus carried to the central station at the Ministry of the Interior.

In Europe, the telegraphic wires have until very lately invariably followed the course of railways; and this circumstance has led some to conclude that, but for the railways, the electric telegraph would be an unprofitable project.

This is, however, a mistake. Independently of the case of the Magnetic Telegraph Company already mentioned, the wires in the United States, where a much greater extent of electric telegraph has been erected and brought into operation than in Europe, do not follow the course of the railways. They are conducted, generally, along the sides of the common coach-roads, and sometimes even through tracts of country where no roads have been made.

It has been contended in Europe that the wires would not be safe unless placed within the railway fences. The reply to this is, that they are found to be safe in the United States, where there is a much less efficient police, even in the neighbourhood of towns, and in most places no police at all. It may be observed, that the same apprehensions of the destructive propensities of the people have been advanced upon first proposing most of the great improvements which have signalised the present age. Thus, when railways were projected, it was objected that mischievous individuals would be continually tearing up the rails, and throwing obstructions on the road, which would render travelling so dangerous that the system would become impracticable.

When gas-lighting was proposed, it was objected that evil-disposed persons would be constantly cutting or breaking the pipes, and thus throwing whole towns into darkness.

Experience, nevertheless, has proved these apprehensions groundless; and certainly the result of the operations on the electric telegraph in the United States goes to establish the total inutility of confining the course of the wires to railways. Those who have been practically conversant with the system both in Europe and in America, go further, and even maintain that the telegraph is subject to less inconvenience, that accidental defects are more easily made good, and that an efficient superintendence is more easily insured on common roads, according to the American system, than on railways.

These reasons, combined with the urgent necessity of extending the Electric Telegraph to places where railways have neither been constructed nor contemplated, have led to the general departure of the telegraphic wires from the lines of railway in various parts of the continent. In France, particularly, almost all the recently constructed telegraphic network is spread over districts not intersected by railways, and even where railways prevail, the wires are often, by preference, carried along the common road.

93. After underground and submarine wires had been constructed and laid upon a considerable scale, the attention of Dr. Faraday was called by some of the parties engaged in their management to peculiar phenomena which had been manifested in the telegraphic operations made upon the lines thus laid. After experiments had been made upon a large scale with lines of subaqueous and subterranean wires, extending to distances varying from

100 to 1,500 miles, it was found that the electricity supplied by the voltaic battery to the covered wire was in great quantity arrested there, by the attraction of electricity of an opposite kind evolved from the water or earth in which the .wire is sunk; the attraction acting through the gutta-percha covering exactly in the same manner as that in which the electricity developed by a common electric machine, and deposited on the inside metallic coating of an electric jar, acts through the glass upon the natural electricity of the external coating, or of the earth in connection with it. The two opposite electricities on the inside and outside of the coating of the wire by their mutual action neutralise each other, and under certain circumstances a person placing his hands in metallic connection with both sides of such coating, may ascertain the presence of a large charge of such neutralised fluid, by receiving the shock which it will give like that of a charged Leyden jar.

The only defect complained of in the underground wires is that which proceeds from accidental failures of complete insulation, produced by defects in the gutta-percha or other coating which allow moisture to penetrate in wet weather and to reach the conducting wire, or it may arise from accidental fracture of the wire. In any such cases the flow of the current to its destination is interrupted, and the telegraph conveys no signal.

94. Although as a general rule the overground lines of telegraphic wire are sustained by supports at intervals of about sixty yards, many exceptional cases are presented in which they are extended between supports at much greater distances asunder. In London, and nearly every large town, great stretches of wires may be seen crossing the streets diagonally between posts several hundreds of yards apart, and in some cases broad rivers, like the Thames, are spanned by the aerial conductors.

Surprising examples of long lines of wires without intermediate support, are presented on the telegraphic line passing north and south through Piedmont, between Turin and Genoa. There, according to a report published in the "Piedmontese Gazette," in the course of the line passing through the district intersected by the chain of the Bochetta, the engineer, M. Bonelli, had the boldness to carry the wires from summit to summit across extensive valleys and ravines, at immense heights above the level of the ground. In many cases the distance between these summits amounted to more than half a mile, and in some to nearly three-quarters of a mile. In passing through towns, this line is carried underground, emerging from which it is again stretched through the air from crest to crest of the Maritime Apennines, after which it finally sinks into the earth, passing through Genoa under the streets, and terminating in the Ducal palace.

It is stated that the insulation of the wires on this picturesque line has been so perfect, notwithstanding the adverse circumstances of its locality, that although it was constantly at work day and night during the first winter, no failure of transmission or extraordinary delay ever occurred.

95. On the introduction of the electric telegraph in India, Dr. O'Shaughnessy of the East India Company's medical department, in constructing an experimental line through a distance of 80 miles from Calcutta, used, instead of wires, iron rods, being the only obtainable materials. These were fastened together and supported on bamboos.

By experiments thus made, he found that the wires employed in Europe would be quite inadequate to the Indian telegraph. In England, where the lines are carried along railways, and where there are no living obstacles to contend with, the thin iron wire, No. 8 or 10 gauge, answers its purpose well; but no sooner were the rods mounted on their bamboo supports in India than flocks of that largest of all birds, the adjutant, found the rods convenient perches, and groups of monkeys congregated upon them; showing clearly enough that the ordinary wire would be insufficient to bear the strains to which these telegraphic lines would be subjected. It was found also that not only must the wire be stronger, but that it must be more elevated, to allow loaded elephants, which march about regardless of roads or telegraphic lines, to pass underneath.

The telegraphic communication thus practically effected, is subjected to attacks to which the telegraphs in this country are but little exposed. Storms of lightning destroyed the galvanometer coils, and hurricanes laid prostrate the posts.

One of the peculiar characteristics of the railway lines intended for India, as contrasted with the English lines, is the greater distance between the posts, which are higher and stronger than those generally used. The thick wire is raised to a height of fourteen feet, on posts nearly the eighth part of a mile apart. To obtain the necessary strength to bear the strain, the posts are fixed with screw piles. To show the strength of the wires thus extended, a rope was, for experiment, hung to the centre- of the wire of largest span, and a soldier climbed up it, the weight of his body producing but a slight curvature. The common deflection arising from the weight of a wire of a furlong span does not exceed eighteen inches.

The plan adopted for joining the lengths of the thick galvanised wire is to have the two ends turned, so as to link into one another, which are then introduced into a mould, like a bullet mould, and an ingot of zinc being cast over them, they form a most substantial joint, and perfect metallic connection. At first it was assumed by the constructors of the Indian telegraphs that, from the general dryness of the climate and thickness of the wire, on long single lines of wire, insulators could be dispensed with; but after ten years of wretchedly bad working on very short circuits, an improved system was adopted, and a better kind of insulation applied to some of the lines. At the present time, however, the internal Indian telegraph is still one of the most insufficient existing.

96. One of the objections against the underground system of conducting wires, was, that while they offered no certain guarantee against the accidental occurrence of faulty points where their insulation might be rendered imperfect, and where, therefore, the current would escape to the earth, they rendered the detection of such faulty points extremely difficult. To ascertain their position required a tedious process of trial to be made from one testing post to another, over an indefinite extent of the line.

A remedy for this serious inconvenience, and a ready and certain method of ascertaining the exact place of such points of fault without leaving the chief, or other station at which the agent may happen to be, has been invented and patented by the Messrs. Bright of the Magnetic Telegraph Company.

Instruments called Galvanometers, which will be more fully described hereafter, are constructed, by which the relative intensity of electric currents is measured by their effect in deflecting a magnetic needle from its position of rest. The currents which most deflect the needle have the greatest strength, and currents which equally deflect it have equal strength.

Electricity always selects the shortest and easiest route to pass by. A thin wire offers more resistance than a thick wire of the same metal, exactly in the ratio of the sectional area of one to the other. Thus a yard of very thin copper wire will offer fifty times as much resistance to the current as a yard of copper wire fifty times its sectional area and weight; therefore one yard of the thin wire will be an electrical measure of fifty yards of the thicker wire, and so on.

The method of Messrs. Bright is founded upon this property of currents. A fine wire, wrapped with silk or cotton so as to insulate it and prevent the lateral escape of the current, is rolled upon a bobbin like a spool of cotton used for needlework. A considerable length of fine wire is thus comprised in a very small bulk. The wire on such a bobbin being connected by one end with the wire conducting a current, and by the other end with the earth, will transmit the current with a certain intensity depending on its length, its thickness, and, in fine, on the conducting power of the metal of which it is made.

Now, let us suppose that a certain length of the far larger wire of the telegraphic line be taken, which will pass a current with the same ease. A galvanometer placed in each circuit will then be equally deflected. But if the length of the line wire be less or greater than the exact equivalent length, its galvanometer will be more or less deflected by it than the other is by the bobbin wire, according as its length is less or greater.

It is, therefore, always possible by trial to ascertain the length of line wire, which will give the current as ready a passage as that which it has upon any proposed bobbin wire.

Bobbins may therefore be evidently made carrying greater or less lengths of wire, equivalent to various lengths of line wire. Suppose then a series of bobbins provided, which in this sense

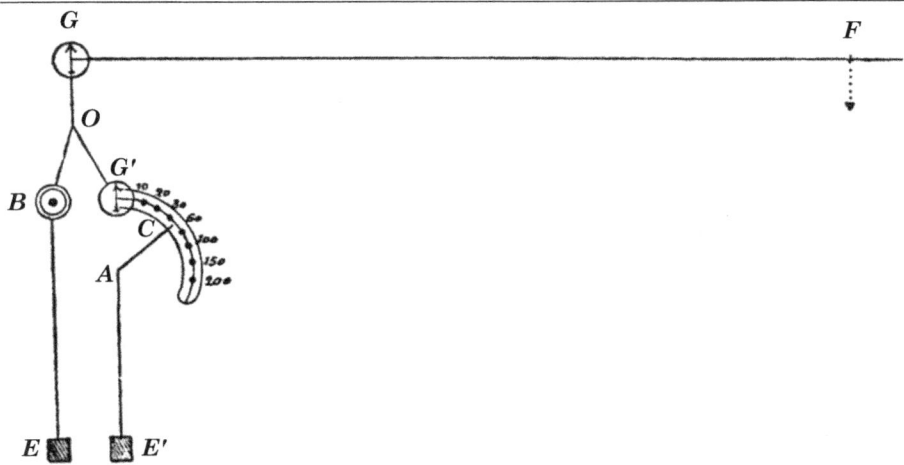

Fig. 44
Bright's galvanometer for detecting faults

represent various lengths of line wire from 100 feet to 300 miles, and let means be provided of placing them in metallic connection in convenient cases.

Such an apparatus is the Messrs. Bright's method of detecting the points of fault, patented by them in 1852. Shown in Figure 44 Let **B** be the station battery, **G** a galvanometer upon the line wire, **F** the point of fault at which the current escapes to the earth, in consequence of an accidental defect of the insulation. Let a wire be attached to the line wire of the station, at **O**, and let it be connected with the first of a series of bobbins such as are described above; let a galvanometer, similar to **G**, be placed upon it at **G'**. Let a metallic arm **A C**, turning on the point **A**, be so placed that its extremity **C** shall move over the series of bobbins, and that by moving it upon the centre **A**, the end **O** may be placed in connection with the wire of any bobbin of the series. Let **A** be connected by a conducting wire with the earth at **E'**, the negative pole of the battery **B** being connected with the earth at **E**.

The apparatus being thus arranged, let us suppose that the wire **A C** is placed in connection with the first bobbin, representing 10 miles of the line wire, and that the distance **G F** of the point of fault is 145 miles. In that case the battery current will be divided at O, between the two wires O G and O G', but the chief part will flow by the shortest and easiest route, and the galvanometer G' will be very much, and G very little, deflected. This will show that F must be very much more than 10 miles from the station. The arm A C will then be turned successively from bobbin to bobbin. When directed to the second bobbin, the current on O G' will have the same intensity as if it flowed on 20 miles of line wire, when turned to the third the same as if it flowed on 30 miles of line wire, and so on. The needle of G' will, therefore, continue to be more deflected than that of G, although the difference will be less and less, as the number of bobbins brought into the circuit is increased. When the bobbins included represent 140 miles, G' will be a little more, and when they represent 150 miles it will be a little less deflected than G, from which it will be inferred that the point of fault lies between the 140th and the 150th mile from the station. A closer approximation may then be made by the introduction of shorter bobbins, and this process may be continued until the place of the fault has been discovered with all the accuracy necessary for practical purposes.

In testing, only one magnetic needle is used, practically; the two wires from O leading to the defective conductor and to the resistance bobbins (or coils) respectively, are wound on a galvanometer coil in opposite directions, so as to throw the needle over in whichever direction the most electricity passes.

Modifications of this apparatus are also employed to show the position of any contact between wires interrupting signals.

This method has been of the greatest practical use in ascertaining the position of faults in the cables to Holland and elsewhere; and in September, 1858, after the Atlantic cable, between Ireland and Newfoundland, so successfully laid in the previous month, began to give signs of impaired insulation, this system was adopted to discover the distance of the fault. The method is now in general use, especially in connection with submarine cables.

PLATE VI

Apparatus employed in laying early submarine cables

6

SUBMARINE TELEGRAPHY

97. SUBMARINE Telegraphy forms a natural sequence to the success attained in insulating underground wires and short river and dock crossings. When experience had shown that neither contact with earth nor water affected the transmission of electricity when the conducting wires were properly protected, the possibility of opening up telegraphic communication across channels and seas became evident.

The first important project of this kind was the connection of the coasts of England and France by a submarine cable, deposited in the bed of the Channel between Dover and Calais. A concession being obtained from the French government on certain conditions, a single conducting wire, invested with a thick coating of gutta-percha, was sunk by means of leaden weights across the Channel, and the extremities being put into connection with telegraphic instruments, messages were transmitted from coast to coast. One of the conditions of the French concession being that this should be effected before September, 1850, this object was attained, but nothing more; for the frail rope was either broken by fishermen, or the action of the waves near the shore constantly rubbing the rope against the rocky bottom, soon wore off the insulating envelope and rendered the cable useless.

The experiment so satisfactorily resolved the doubts which

had been entertained as to the possibility of sufficiently insulating a wire for any considerable length under water, that it was immediately determined to resort to means for the effectual protection of the conducting wires from the effects of all the vicissitudes to which they would be exposed. The construction of the present Dover and Calais cable was accordingly commenced at Millwall, shortly after the failure of the experimental line. The principle resorted to, suggested by Mr. Küper, was simply that of a colliery rope, the outside iron wires being laid spirally around the gutta-percha covered conducting wires, instead of over the usual hemp core; by this means great strength was combined with an armour protection for the insulated wires within. As a proof of the value of the design, every subsequent cable that has proved a permanent success has been formed upon the plan of this Dover and Calais rope. Improvements have been introduced into the details of many cables, but the principle of spiral wire covering is always more or less adhered to. The cable was coiled on board H.M.S. *Blazer* in September, 1851, and successfully laid under the direction of Messrs. Crampton and Wollaston, the engineers of the Submarine Telegraph Company, an association formed under the auspices of Mr. J. W. Brett and Sir James Carmichael, to carry into effect this enterprise, perhaps the most important inauguration of a new servant of mankind which the world has seen during the present century.

Notwithstanding the enormous traffic up and down the Channel, this cable has been seldom injured during upwards of fifteen years' service, and has been easily repaired on each occasion. It is now (January, 1867) in a perfect state of insulation as regards the whole of its four conducting wires.

98. The success attending the Dover and Calais cable led to the execution of further works of the kind, to connect up England with Ireland, Belgium, Holland, Hanover, and Denmark, and subsequently for the Mediterranean and other seas and channels. In all, there have been no less than seventy-four important cables constructed in this country alone, besides many minor lengths for river crossings.

These cables have been principally constructed by Messrs.

Glass, Elliot, and Co., of Greenwich, Messrs. E. S. Newall, and Co., of Gateshead, and W. T. Henley, of North Woolwich. The inner wires were insulated at the Gutta Percha Company's Works in the City Road, under the direction of the late Mr. Statham and Mr. Chatterton. The works of the Gutta Percha Company have been recently combined with those of Messrs. Glass, Elliot, and Co., under the title of the Telegraph Construction and Maintenance Co., by whom the Atlantic Cables of 1865 and 1866 were made and successfully completed last year between Ireland and Newfoundland.

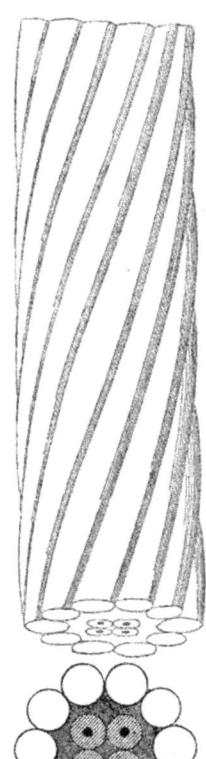

The following table (pp. 102 and 103) shows at a glance the progress which has been made in submarine telegraphy to the present time, January, 1867.

99. In the Dover and Calais cable, which was the first fabricated and laid, each of the four copper wires is surrounded by gutta-percha, which in Figure 45 (lower section) is indicated by the light shading round the black central spot, representing the section of the copper wire. The four wires thus prepared were then enveloped in the general mass of prepared spun yarn, represented by the darker shading. The ten galvanised iron wires were then twisted around the whole, so as to form a complete and close armour.

The external form and appearance of this helical coating is represented in Figure 45 (upper). This cable, which was completed in three weeks, measured originally 24 miles in length. Owing to the manner in which it was laid down, the machinery being naturally crude in design, and the cable hands unaccustomed to their work, it was found insufficient to extend from coast to coast, although the direct distance is only 21 miles. It was therefore found necessary to manufacture an additional mile of cable, which being spliced on to the part

Fig. 45
Dover and Calais
Cable

(Continued on page 104)

LIST OF SUBMARINE CABLES

	From	To	Year when laid.	No. of conducting wires.	No. of iron wires forming outer casing.	Total length, statute miles.	Weight per mile. tons.	cwt.	REMARKS.
1.	Dover	Cape Grisnez (France)	1850	1	none	30	0	4	Experimental. Thick gutta percha wire. Failed day after laying.
2.	Dover	Calais	1851	4	10	25	7	0	Still working well.
3.	Holyhead	Howth (Ireland)	1852	6	12	73	7	0	Failed second day after laid.
4.	Dover	Ostend	1852	6	12	80	7	0	Still working well.
5.	Portpatrick	Donaghadee (Ireland)	1853	6	12	25	7	0	Do.
6.	Denmark	Island of Zealand (across Great Belt)	,,	3	9	18	4	0	Too light for shallow water. Repeatedly broken by anchors and trawlers. Removed in 1859, and replaced by heavier cable.
7.	Orfordness	Hague	,,	1	10	119	2	0	
8.	Orfordness	Hague	,,	1	10	118	2	0	
9.	Orfordness	Hague	,,	1	10	123	2	0	
10.	Firth of Forth		,,	4	10	5	8	0	Composed of 4 cables of 2 tons each, laid together.
11.	Corsica	Sardinia	1854	6	12	13	8	0	Still working well.
12.	Spezzia	Corsica	,,	6	10	110	8	0	Do.
13.	Holyhead	Howth	,,	1	10	73	2	0	Failed upon laying; too light.
14.	Holyhead	Isle of Wight	,,	1	10	73	2	0	Do.
15.	Portsmouth	Whitehead (Ireland)	,,	1	12	1	5	0	Still working well.
16.	Portpatrick		,,	6	12	26	2	0	Do.
17.	Denmark	Sweden	1855	3	10	13	6	0	Do.
18.	Varna	Balaclava	,,	1	none	305	0	15	Used for a short time during Crimean War; then failed.
19.	Balaclava	Eupatoria	,,	1	12	30	0	15	
20.	Varna	Constantinople	,,	1	12	171	2	0	
21.	Orfordness	Hague	,,	1	10	119	2	0	
22.	Newfoundland	Cape Breton Island	1856	3	12	85	2	0	Too light; removed.
23.	Cape Breton Island	Nova Scotia	,,	1	12	1¼	4	0	Still working; repaired 1866
24.	Prince Edward's Island	New Brunswick	,,	4	18	12	2	0	Failed 1864.
25.	Sardinia	Algeria	1857	1	18	178	3	0	
26.	Sardinia	Malta	,,	1	18	365	0	18	Failed after several years.
27.	Malta	Corfu	,,	1	18	390	0	18	Failed after a year; subsequently laid to Sicily.
28.	Norway	Across Fiords	,,	1	18	49	0	18	Failed after 3 years.
29.	Dardanelles	Scio	1858	1	18	187	2	0	
30.	Scio	Candia	,,	1	18	330	1	0	Failed after several years.
31.	Ceylon	Hindostan	,,	1	10	30	1	0	Do.
32.	Australia	King's Island	,,	1	10	140	2	0	Still working.
33.	Valencia (Ireland)	Newfoundland	,,	1	18	2200	1	0	Failed after a month, through defective insulation.

No. & From	To	Year			Miles			Remarks
34. Orfordness	Haerlem (Holland)	..	4	10	156	9	15	Still working well.
35. *Weymouth*	*Channel Islands*	..	1	9	93	2	17 {	Too light for situation; broken repeatedly. Abandoned in 1860.
36. Whitehaven	Isle of Man	1859	1	10	36	2	10	Still working. Repaired.
37. *Cromer*	*Hanover*	..	2	12	280	3	0	Failed 1863; too light for situation.
38. *Liverpool*	*Holyhead*	..	2	18	25	3	0	Do. 1862. Do. Do.
39. *Athens*	*Syra*	..	1	18	70	1	0	Do. after several years' working.
40. Syra	Scio	..	1	18	108	1	0	
41. Cromer	Heligoland and Denmark	..	3	12	359	4	0	Broken 1865.
42. Folkestone	Boulogne	..	6	12	25	9	10	Still working well.
43. *Singapore*	*Batavia*	..	1	18	530	1	0	Failed soon after laying; too light.
44. Sweden	Gothland	..	1	12	64	2	10	Still working.
45. King's Island (Bass's Straits)	Tasmania	..	1	12	100	2	0	Do.
46. Sicily	Malta	..	1	10	70	3	0	Do.
47. *Jersey*	*Pirou (France)*	1860	1	10	21	3	10	Do.
48. *Suez (Red Sea, via Aden)*	*Kurrachee (India)*	..	6	18	3500	8	0	Partly laid 1859; failed shortly after.
49. Denmark	Great Belt	..	1	12	14	5	10	Still working.
50. Denmark	Great Belt	..	3	18	14	1	0	Do.
51. Dacca	Pegu	..	1	18	216	1	0	
52. *Barcelona*	*Port Mahou (Minorca)*	..	2	16	180	1	5	Failed 1865.
53. Minorca	Majorca	..	1	18	35	1	18	Still working.
54. Majorca	Iviza	..	2	18	74	1	18	Do.
55. Iviza	St. Antonio (Spain)	..	2	18	76	1	18	Do.
56. Minorca	Algiers	1861	1	10	239	2	3	
57. Toulon	Corsica	..	1	10	195	2	3	Do.
58. *Holyhead*	*Howth*	..	1	10	76	2	0	Failed 1865. Too light for channel.
59. Corfu	Otranto	..	6	18	60	3	9	Still working.
60. Malta	Alexandria	1862	6	18	1535	8	0	Do.
61. Beachy Head	Dieppe	..	4	12	63	6	0	Do.
62. Pembroke (South Wales)	Wexford	1863	4	12	130	4	0	Do.
63. Lowestoft	Holland	1864	3	12	3	3	10	
64. Ireland	Cape Clear Island	..	1	10	1430	4	10	Do.
65. Persian Gulf (Fao)	Kurrachee	1865	1	12	66	3	0	Do.
66. Otranto	Avlona	..	1	12	97	1	10	Do.
67. La Calle (Algeria)	Biserta (Tunis)	1866	1	12	165	1	10	Do.
68. Biserta	Marsala (Sicily)	..	3	9	55	4	10	
69. Sweden	Prussia	..	1	12	25	2	10	Do.
70. Portpatrick	Antrim	..	1	10	1950	1	11	Laid July, 1866.
71. Valentia (Ireland)	Newfoundland	..	4	10	1960	1	15 {	1,200 miles laid in 1865; end broke. Recovered and remainder laid, Sept., 1866.
72. Valentia	Newfoundland	..	4	12	240	6	0	
73. Lowestoft	Norderney (Hanover)	..	1	10	86	2	15	
74. Newfoundland	Cape Breton Island							

Total mileage . . . 19,923

103

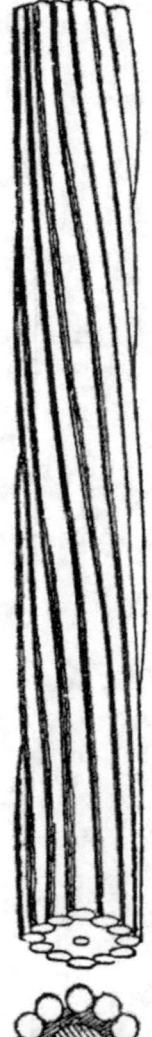

(Continued from page 101)

laid, the whole was completed, and the electric communication between Dover and Calais definitively established on the 17th October, 1831.

The cost of the cable itself was 9000*l*, being at the rate of 360*l* per mile.

100. An attempt, which proved a failure, was then made to connect Holyhead on the Welsh with Howth on the Irish coast. While several companies which had been formed for the purpose, were occupied in raising the capital necessary for this project, they were surprised by the announcement that the project was already on the point of being realised by Messrs. Newall and Co., on their own account.

The distance between the points to be connected being 60 miles, the cable was made with a length of 10 additional miles, to meet contingencies. In this cable, which enclosed only one conducting wire, the external wires enclosing the insulating rope were made thicker at the parts near the shores than for that which lies in deep water, the former being subject to much greater disturbing forces. A side view of the part immersed in deep water, and its cross-section, is given in Figure 46. A side view of the shore ends, with a corresponding cross-section, is given in Figure 47.

The gutta-percha rope was fabricated by the Gutta Percha Company,

Fig. 46
Holyhead
and Howth
deep sea part

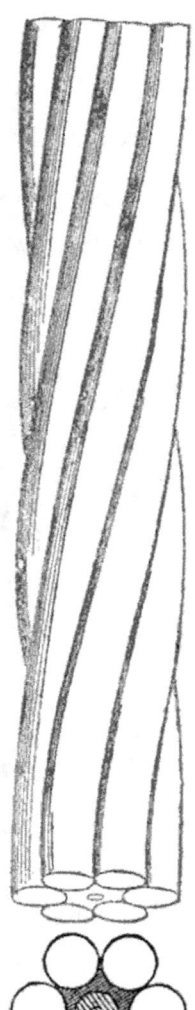

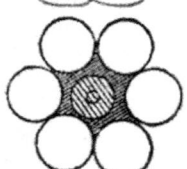

Fig. 47
Holyhead and
Howth shore
ends

London, from whence it was sent to Gateshead, where it received the iron wire envelope at the works of Messrs. Newall and Co., in the short space of four weeks. Loaded on twenty wagons, it was next sent by railway across England to Maryport, where it was embarked on board the *Britannia*, and transported to Holyhead.

On the 1st June, 1852, it was laid in the bed of the channel. The depth of water is 60 fathoms, being more than twice that of Dover.

The entire process of laying it down was completed in 18 hours. In another hour the cable was brought ashore, and put in connection with the telegraphic wires between Howth and Dublin, and immediately afterwards London and Dublin were connected by means of instantaneous communication.

This cable was lighter considerably then that between Dover and Calais, its weight being a little less than one ton per mile, and consequently its total weight did not exceed 80 tons; while the Dover and Calais cable weighing 7 tons per mile, its total weight was 180 tons.

From some cause this cable, after being worked for three days, became imperfect. Its strength was wholly miscalculated for the Irish Channel, where the tides are strong, bottom irregular, much anchorage, and frequent trawling. None but the heaviest class of cables have been found to answer in such situations, but this attempt must be regarded in the light of an experiment.

On the 9th October, 1852, Messrs. Newall and Co. attempted to lay a cable across the narrowest part of the Irish Channel between Portpatrick and Donaghadee for the Magnetic Telegraph Co. This cable contained six conducting wires similar to Figure 48. The distance across is the same as between Dover and Calais, viz., 21 miles, and 25 miles of cable were placed on board the *Britannia* steamer. The process of submersion was carried on until 16 miles had been successfully laid down, when a sudden gale came on, which rendered it impossible to steer the vessel in the proper course, and Mr. Newall was reluctantly compelled to cut the cable, when within 7 miles of the Irish coast, and having 9 miles of cable remaining on board.

The whole of this 16 miles of cable was recovered in June,

1854, after being nearly two years submerged. This proved an arduous undertaking. The depth of water in this part of the Irish channel is 150 fathoms, or 900 feet, and from this depth the cable was dragged by means of a powerful apparatus worked by a steam engine placed on the deck of a steamer. The operation occupied four days, for from the great force of the tide, which runs at the rate of 6 miles an hour, it was found impossible to work except at the times of high and low water. The cable was also imbedded in sand, so that the strain required to drag it up was occasionally very great. The cable was in perfect order when recovered.

101. The next great enterprise of this kind was the deposition, by the Submarine Telegraph Company, in the bed of the Channel, of a like cable connecting the coasts of England and Belgium, measuring 70 miles. This colossal rope of metal and gutta-percha was also constructed by Messrs. Newall.

The extension of these extraordinary media of social, commercial, and political communication between countries separated by arms of the sea, may be conceived, when it is stated that during the past 16 years no less than 19,923 miles of submarine cable have been made and laid beneath the sea.

The cable laid between Dover and Calais includes, as already stated, four conducting wires. That between Dover and Ostend contains six wires insulated by the double covering of gutta-percha, manufactured, under Mr. S. Statham's directions, by the Gutta Percha Company. The gutta-percha laid into a rope is served with prepared spun-yarn, and covered with twelve thick iron wires, of a united strength equal to a strain of 40 to 50 tons — more than the proof strain of the chain cable of a first rate man-of-war. A side view and section of this cable is given in Figure 48.

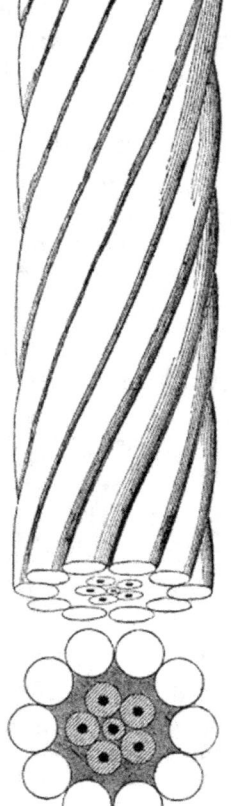

Fig. 48
Dover and Ostend

The Belgian cable weighed 7 tons per mile, so that its total weight was about 500 tons. Its cost was 33,000*l.* It took 100 days to make it, 70 hours to coil it into the vessel from which it was let down into the sea, and 18 hours to submerge it.

The cable was stowed away in the hold in an oblong coil of considerable length. On the morning of the Wednesday, the 4th May, 1853, the screw steamer *William Hutt*, Capt. Palmer, freighted with the cable, being anchored off Dover, near St. Margaret's, South Foreland, the process of laying the cable was commenced. This vessel was attended and aided by H.M.S. *Lizard*, Capt. Rickets, R.N., and H.M.S. *Vivid*, Capt. Smithett. Capt. Washington, R.N., was appointed, on the part of the Admiralty, to mark out the line and direct the expedition.

At dawn of day about 200 yards of the cable were given out from the *Hutt*, and were extended by small boats to the shore, where the extremity was deposited in a cave at the foot of the cliff. There telegraphic instruments were provided by means of which, through the cable itself, a constant communication with the vessel was maintained, corresponding telegraphic instruments being placed on board the *Hutt*.

The arrangements adopted for paying out these strong cables in shallow water are shown in Plate VI, (p. 98); the cable as it came up from the hold, being passed several times round a large brake-wheel, by means of which the cable was kept from going out too fast, and its motion maintained so as to be equal to the progress of the vessel. Men are represented in the figure applying the brake to the wheel.

On arriving off Middlekerke, on the Belgian coast, a boat sent from shore took from 500 to 700 yards of the cable on board, for the purpose of landing it. The boats of the British vessels taking her in tow, the end of the cable was safely landed, and deposited in a guard-house of the Custom House, where the telegraphic instruments brought in the *Hutt* being erected, and the communications made, the following despatch was transmitted direct to London:–

Union of Belgium and England, twenty minutes before one, p.m. 6th May, 1853.

102. The next submarine cable laid in May 1853 was that of the Magnetic Telegraph Company, connecting Donaghadee with Portpatrick, also manufactured by Messrs. Newall and Co.

This cable, which contains six conducting wires, is represented in Figure 49, and corresponds in weight and form to the Belgian cable. But in the details of its construction and composition, some improvements were introduced. This rope was manufactured in 24 days, and cost about 13,000*l.*

The cable laid down in 1854 by the British Telegraph Company between the same points, is precisely similar to this.

103. Orfordness, on the Suffolk coast, was subsequently connected with the Hague, by four separate submarine cables, each containing a single wire. Near the shore on each side these were brought together and twisted into a heavy cable, as represented in Figure 50. The distance from Orfordness to the Hague being 120 miles, the cables were made 135 miles in length. They were laid down separately at a little distance one from another. At 3½ miles from the shore they were brought together. It has been found, however, that these separate light cables are subject to frequent injury from anchorage, and the Electric Company to which they belonged replaced

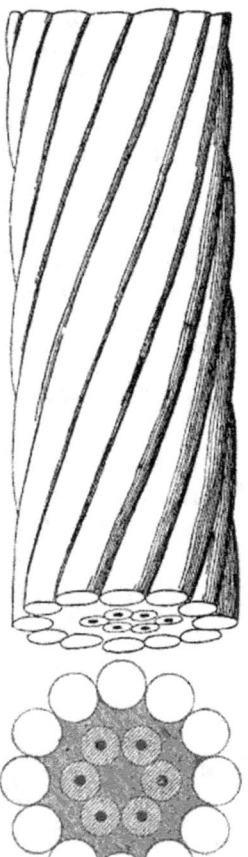

Fig. 49
Donaghadee and
Portpatrick

them in August, 1858, by a much stronger compound cable, made and laid for them by Messrs. Glass, Elliot, and Co., — weight, 9 tons per mile.

104. Following the submersion of the submarine lines enumerated, a heavy cable containing six wires was successfully laid in the Mediterranean in June, 1854, by Mr. Brett, from Spezzra, in Piedmont, to the northern coast of the Island of Corsica,

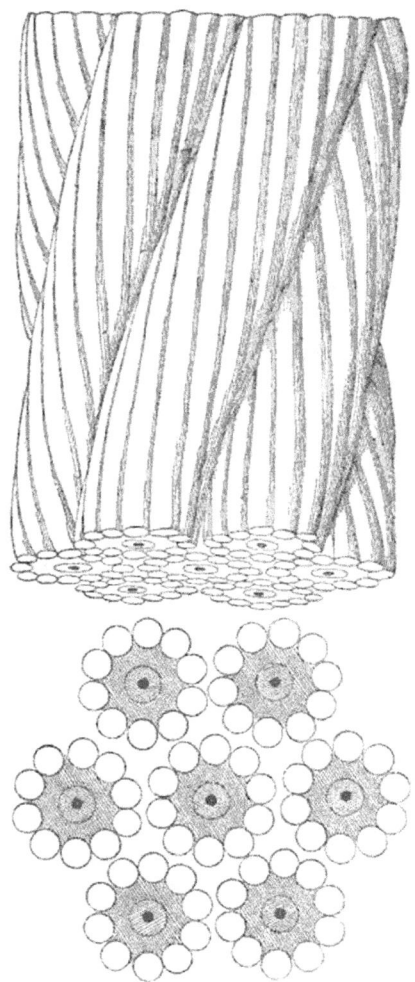

Fig. 50
Orfordness and the Hague

a distance of 100 miles. The cable was one similar to that represented in Figure 48.

In this expedition far greater depths of water had to be passed over than had previously been attempted with submarine cables, the soundings at some points indicating 450 fathoms, or upwards of half a statute mile between the surface of the water and the bottom.

During the process of submersion, and after it was believed that the deepest soundings had been safely surmounted, the cable slipped upon the surface of the brake-drum used to check it, and flew out from the vessel with great velocity, cutting the bulwarks in its passage. As soon as the career of the cable had been stopped, it was found that a portion of the length which had surged over the drum was so squeezed and flattened, that its electrical condition was defective. With some difficulty the injured part was drawn back into the vessel and repaired, after which the remainder was safely laid. A short piece of cable was extended at the same time between Corsica and Sardinia.

This line has continued in successful operation ever since its completion, thus forming a section of the telegraphic communication established between France and her possessions in Algeria, and between this country and Malta and Corfu.

Two attempts were made by Mr. Brett in the two following

years to connect Cagliari on the south coast of Sardinia with Cape Bona in Algeria, but failed from the length of cable on board falling short in each case. Soundings of nearly two miles depth had to be passed over, owing to which it appears that one, if not both, of these cables payed out occasionally too quickly. In the latter attempt the cable was safely submerged to within a dozen miles of the African coast, a distance of 135 miles, when it fell short. A message was sent through the cable to Greenwich for an additional length, and the vessel held on by the cable for five days, until it parted at last in 500 fathoms, owing to friction on the bottom. This cable weighed four tons per mile, and shortly after starting with it, a slip took place on the paying-out drum, and the cable broke, but the end was recovered again after under-running it for eighteen miles.

105. The short submarine cable laid down between Prince Edward's Island, and the coast of Nova Scotia (Figure 51), is part of a more extended submarine line connecting Newfoundland with Canada. The other sections make up a total length of 140 miles.

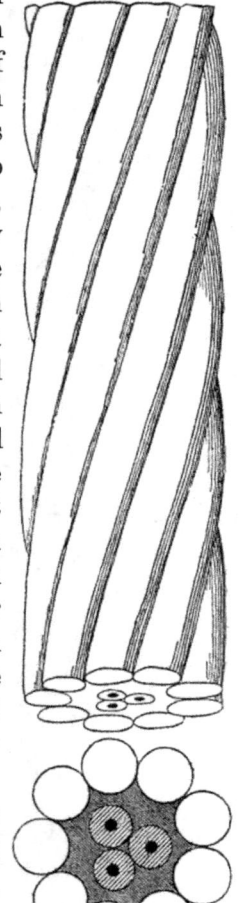

Fig. 52
Great Belt

Fig. 51
P. Edward's Island and N. Brunswick

The Legislative bodies of Newfoundland, Nova Scotia, and Maine, have granted exclusive privileges to the company that carries out these extensions, including a considerable grant of land. These privileges have been subsequently combined with the Atlantic Telegraph undertaking, which is now worked in connection with these lines from New-

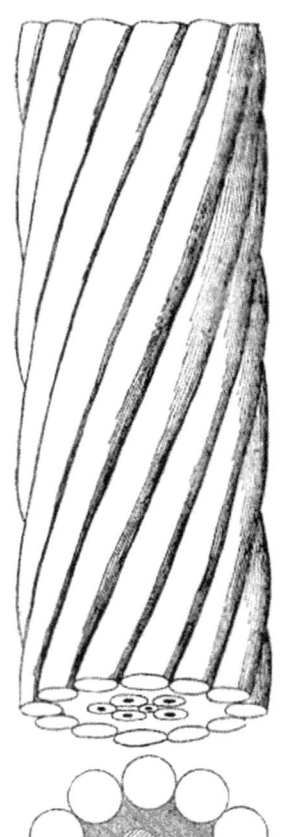

Fig. 53
Zuyder Zee

foundland to Portland, in Maine.

106. The Danish submarine cable, Figure 52, is carried across the Great Belt from Nyborg to Korsoe, the nearest point of the opposite coast of Zealand.

The cable laid across the Zuyder Zee is shown in Figure 53.

Subaqueous cables have been laid across several of the American rivers. The difficulties supposed to attend the deposition and preservation of these conductors appeared to telegraphic engineers and projectors so formidable, that the wires were at first carried across the rivers between the summits of lofty masts erected on their banks. This method, however, was found to be attended with such effects as to render the maintenance of the wire impracticable. The masts were blown down by the violent storms and tornadoes incidental to the climate, and were not infrequently destroyed by lightning.

The project of depositing the conducting wires in the bottom of the river was then resorted to, and has been carried into effect in several cases.

107. The cable between Cape Breton and Newfoundland, a distance of 74 miles, was manufactured and laid by Messrs. Glass, Elliot, and Co., during the summer of 1856. An improvement was introduced in the conductor of this cable, which consists of seven small copper wires twisted together into a strand, with the view to prevent any flaw in a copper wire at any point stopping the conductivity of the rest. This cable, Figure 54, formed part of the line of .communication between England and the United States, during the temporary working of the Atlantic Cable in August and September, 1858.

At the end of 1857 light single wire cables, protected by an external casing of small iron wires, were laid by Messrs. Newall for the Mediterranean Extension Telegraph Company, between Sardinia, Malta, and Corfu, and were followed by a line to the Channel Islands from Weymouth.

Cables to Emden, in Hanover (280 miles), and to Heligoland, and thence to Tonning, in Denmark (360 miles), were successfully submerged in October, 1858, and July, 1859, respectively, for the Submarine Telegraph Company, by Messrs. Glass, Elliot and Co., from Weybourne, in Norfolk, where the wires are connected to the land lines of the Magnetic Company. The Hanoverian contains two conducting wires, and weighs three tons per mile; the Danish three wires, and weighs three and a half tons per mile.

These cables have, however, proved too light for the comparatively shallow water in which they are laid for a considerable distance, and have been frequently injured by anchors and by fishermen when trawling.

The telegraphic intercommunication increased so rapidly between France and England, that in 1859 an additional cable of great strength, containing six wires, was laid between Folkestone and Boulogne, by Messrs. Glass and Elliot, for the Submarine Telegraph Company; and again in 1861 another cable, also containing six wires, was laid for the same company between Beachy Head and Dieppe.

108. Other countries are vying with England in carrying out submarine works. In 1860 the Spanish Government arranged with Sir Charles Bright and Mr. Henley to open up correspondence by telegraph with Port Mahon, and connect the Islands of Majorca, Minorca, and Iviza with the mainland both at Barcelona and St. Antonio. These cables were successfully submerged in very great depths — that between Barcelona and Port Mahon being

Fig. 54
Cape Breton and New-foundland

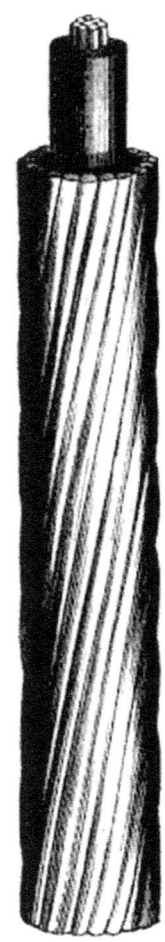

Fig. 55
Red Sea
1860

1400 fathoms, or more than a mile and a half.

A cable 240 miles long, containing a single conductor, has been laid by Mr. Henley between Tasmania and Australia, to put the Van Diemenites *en rapport* with their fellow colonists.

109. 3500 miles of light single wire cable were laid in sections, in 1859 and the beginning of 1860, between Suez, Aden, and Kurrachee, by Messrs. Newall, for the Red Sea and India Telegraph Company; but the various sections broke down before they could be opened throughout for public use. This great failure is ascribed to the cable being much too fragile in construction, and laid far too taut over an uneven bottom. The want of slack to fill up the inequalities, and the weakness of the line, made it give way shortly after being laid, and prevented this, the best telegraphic route to India, from being brought into use. This cable is shown in Figure 55.

110. In 1861 a cable was laid for Government between Malta and Alexandria, by Messrs. Glass and Elliot. The distance is 1535 miles, and the line touches at two points on the African coast, which are used when necessary as transmitting and repeating stations. This cable has continued to work most satisfactorily, and a very large business has been developed between Egypt and various places in England and Europe, to which this line gives access. It was originally intended by Government to form a connection between Falmouth and Gibraltar, and was made upon a specification recommended by Sir Charles Bright to Government in a report to Sir Stafford Northcote, in 1859.

111. The history of submarine telegraphy has been one of frequent success, but also of frequent failure, arising either from imperfect insulation, cables ill-adapted for the work, or laid too taut over uneven bottoms, or else through the paying-out machinery being inadequate for the requirements of the work. These points are better understood now, and permanent success takes the place of disaster.

Some eminent scientific authorities express doubts as to the durability of the submarine cables. In the case of the Dover and Calais cable it has been observed that the bottom of the Channel at that part of the strait is proved by the soundings to be subject to undulations, so considerable that the summits of some of its elevated points rise to such a height that the water which covers them is not deep enough to secure them from the effects of the tumultuous agitation of the surface in violent storms. It is here well to remind the reader that the agitation of the ocean, which seems so awful in great tempests, has been found to extend to a very limited depth, below which the waters are in a state of the most profound repose. The objection we now advert to is, therefore, founded upon the supposition that the crests of some of the elevations upon which the submarine cable rests are so raised as to be within that limit of depth, and it is feared that, such being the case, the violence of the water in great tempests may so move the cable against the ground on which it is deposited with a motion to and fro, as to wear away by frequent friction its metallic armour, and thus expose the conducting wires within it to the contact of the water, and destroy their insulation.

But it has been most satisfactorily proved by a part of the experimental wire which was laid down between Dover and Calais, in 1850, and which was picked up two years afterwards in as perfect a state as when laid down, that the action of the waves does not affect the bottom of the Channel there. The greatest depth is 30 fathoms, and the bottom shelves regularly from Dover to near Cape Grinez, where there is a ledge of rocks rising suddenly from the bottom.

It has been also feared that, notwithstanding the effect of the galvanisation of the surface of the surrounding wires, the corrosive action of the sea water may in time destroy them; and it has been suggested that some better expedient for protection against this effect might be contrived upon the principle suggested by Davy, for the preservation of the copper sheathing of ships, by investing the cable at certain intervals with a thick coating or glove of zinc, which would increase the efficiency of the thinner coating of that metal given to it in the process of galvanisation.

To this practical men who have had as much experience as is compatible with the recent date of these novel and extraordinary enterprises, reply that the results of their observations give no ground for apprehension of any injurious effects from tidal or tempestuous action, and that the fine iron used in the wire is not affected by seawater, as larger masses of coarser iron, such as anchors, are. They cite as proof of this, the slightly decayed state in which nails and small fire-arms have been found when recovered from vessels long sunk. They further state that the tar contained in the layer of hemp within the protecting wires acts as a preservative, whether the wires be galvanised or not. It has been found for example that, in the case of the submarine conductor between Donaghadee and Portpatrick, a perfect concrete of tar and sand was formed, upon which masses of shell-fish attached themselves at all parts that are not buried in sand, and in a few years a calcareous deposit was formed around it, which cemented it to the bottom, and altogether intercepted the action of the sea water.

112. In the deposition of submarine cables great care should be taken to select suitable points on the shore for beaching them. Sandy places are always to be sought. If this precaution be taken, it is affirmed that they are not subject to tidal action in shallow water. A cable was partly laid for the Magnetic Telegraph Company in 1852 near Portpatrick (102), but abandoned in consequence of the vessel employed to deposit it being exposed in the process to a violent storm. The wire was left exposed upon the beach down to and beyond low water mark, and was in June, 1854, still in a perfect state, the galvanised iron wires, even to their zinc coating, being absolutely in the same state as when they were deposited.

It is contended by practical men that the great and only risk of failure in the submarine cables is from defects produced in the process of their deposition, or from original faults in the principle of their construction.

The greatest care is necessary in conducting the process of delivering out the cable into the sea, or "paying it out," as it is technically called. All sudden bending of the cable is to be especially avoided. "Kinks" or "hitches" were apt to occur in the process, by which the gutta-percha covered wires within the cable become strained. Some of the early cables are found, when repairs are made at various points, to have had many serious kinks during laying; but by the apparatus now used in the holds of vessels paying out cables, such injuries are guarded against.

It is said that the Belgian cable has been subject to some imperfection arising from the position of the wires within the case. The sixth wire being in the axis of the cable, surrounded by the other five, it was found that when the outer casing of the protecting wires was laid around it, the pressure on the centre wire rendered it imperfect, while the five surrounding it suffered to some extent.

Similar defects are said to exist in other cables constructed upon the same principle.

A hempen case well tarred in the centre is considered to form the best safeguard for the gutta-percha covered wires in the process of making the cable, since it will yield to any compression itself without affecting injuriously the wire. It is also found desirable to give the outer wires a wrapping of hemp saturated with a compound of pitch and tar, or other preservative material, to prevent rusting.

The results shown by the tables on pages 102 and 103, may be thus summed up:–

The total length of submarine cables (exclusive of short lengths for river crossings) laid during the 16½ years between June, 1850, and the end of 1866, amounted to 19,923 miles of cable, containing 25,653 miles of conducting wire — many being multiple, with three, four, or six conductors.

Of this vast length, no less than 9,303 miles of cable, or nearly one-half of the whole, and comprising 10,146 miles of conducting wire, have failed, and been abandoned. The Atlantic cable of 1858, and the Red Sea and India line of 1859-60, contributed 5,709 miles of this loss between them.

The entire capital that has been invested, up to the present time, in making and laying the 74 submarine cables enumerated, amounts to about 5,500,000*l.*; of which 23 cables, representing 2,200,000*l.*, have been lost. The first Atlantic cable and the Red Sea line made up 1,250,000*l.* of this disastrous total.

If the list be referred to, it will be seen that the costly experience thus purchased has been derived from the failure of a series of very light and weak cables; insufficient either in strength or insulation for the positions in which they were placed, or the work required of them.

On the other hand, the strong well-protected cables all last; and their comparative durability, as contrasted with their frailer brethren cords, may be estimated from the fact that, without any exception, all the strongly-armoured ropes exceeding four tons weight per mile are still working, including no less than nine laid between 1851 and the end of 1854; while every light cable laid up to 1859 has already given way.

PLATE VII

The Indo-European Telegraph
Landing the cable in the mud at Fao, Persian Gulf

7

THE TELEGRAPH TO INDIA

113. Failure of the Red Sea line laid in 1859-1860, and survey of the Persian Gulf—114. Description of improved cable designed by Sir C. Bright and Mr. L. Clark—115. Searching tests applied to the gutta-percha wire and joints by accumulation—116. Protection of iron sheathing by yarn and a silica composition—117. Laying the cable; difficulties surmounted—118. Bad working of the land lines in connection with the cable—119. Proposed extension to China and Australia.

113. THE crowning achievements of telegraphy may fairly be said to consist in the connection of Great Britain on the one hand with her vast Indian Empire, and on the other with the American continent.

The latter, and greatest, work will he reserved for the two next chapters, concluding the present record of submarine enterprise. Political, apart from commercial, necessity rendered it imperative that the governments in England and India should be brought within the shortest possible period of communication with one another. It was felt, especially after the Indian mutiny, that in this era of the telegraph the countries could never be allowed to be separated by thirty days of postal service, when by the agency of the wires but a few hours need divide them. Had electric communication existed at the time of the outbreak, it would have been quelled far more speedily, and priceless lives saved; if, indeed, the knowledge among the Sepoys of the existence of this speedy means of summoning aid had not wholly or partially sufficed to deter them from their rising.

With the view of establishing this communication, Government warmly supported and subsidised a company formed in 1858, to lay cables along the Red Sea and the southern shores of Arabia, between Suez, Aden, and Kurrachee. The failure of the attempt to establish the telegraph by this route will be fresh in the memory of all who have paid attention to the progress of the electric lines. Sheathed with a covering of extremely slight wires quite unprotected from corrosion, the various sections of cable were laid by Messrs. Newall with, apparently, little or no allowance for slack, and consequently so taut as to hang from point to point along the coral reefs and inequalities of the bottom. No sufficient care seems to have been exercised even in the selection of the route along the Red Sea; and consequently the cable gave way, and became defective at various points within a few weeks of the laying of each section; thus the line, after costing about 800,000*l.* was never available throughout.

Baffled in their object of effecting communication by this line, and discouraged by the heavy annual loss under the guarantee given upon the capital, our government deferred for a few years further movement in the matter. While the Red Sea line was in course of construction, the Turkish Government were busily engaged in erecting their telegraph from Constantinople across Asia Minor to Diarbekr, and thence by Mosul, the ancient Nineveh, to Bagdad. This line, and its extension to the Persian Gulf, had been urged for a long time by her Majesty's Government, who appreciated the importance of two distinct telegraphs to India. Upon the Bagdad wire being finished the late Colonel Patrick Stewart, R.E., was requested by the Indian Government, in 1861, to survey the coast of the Persian Gulf for an overground telegraph along its shores, to connect the Indian and Turkish telegraphs. Mr. Latimer Clark, C.E., at the same time made a careful examination as to the state of the cable between Suez, Aden, and Kurrachee.

The result of Mr. Clark's investigation showed the impossibility of restoring the Red Sea and Arabian line, and Colonel Stewart reported against the reliability of a land line along the coast of the gulf.

114. Special soundings were made along the Persian Gulf, and the bottom being favourable for a cable, Government resolved to lay one of great strength and durability, to be constructed upon a plan designed by Sir Charles Bright and Mr. Clark, and appointed them engineers to the work.

The total length of cable required was 1450 miles, weighing

no less than 5028 tons, forming by far the heaviest length ever carried by a submarine telegraph expedition. The following description of the manufacture of the cable will give an idea of the improvements introduced. We will first deal with the core. The copper conducting wire weighed 225 lbs. per mile, and was built up of four segmental bars put within a hollow tube about 12 feet long and an inch in diameter, and the whole rolled and drawn down together into a wire of the requisite size, apparently solid, though in reality multiple in construction, and consisting of five

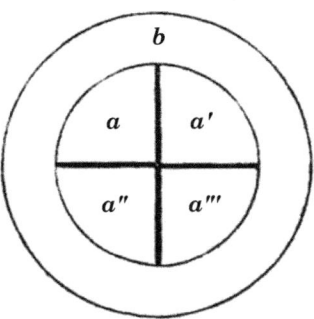

Fig. 56
Diagram of built-up, segmental, conducting wire of Persian Gulf cable.

distinct pieces. By this device the mechanical benefits of a strand wire were combined with the small surface and electrical advantages of a solid one, which admits of a higher speed of conveying messages than the stranded form of conductor. This is shown in section, Figure 56, where *a*, *a'*, *a"*, *a'''*, are the segmental bars and *b* the enclosing tube.

The copper selected to form the conductor, so built up, was specially tested and chosen for its high capacity of conducting electricity. The importance of this will be recognised, as it is proved that samples of copper some times vary in conductivity as much as 40 or 50 per cent, from one another; that is, some specimens will carry electricity with as great facility as others of double the thickness, though physically there may be nothing to show the difference.

In testing the conductivity, the resistance of the copper was determined by means of a Wheatstone's electrical balance. The standard being pure galvano-plastic copper at 100, the lowest conductivity of any of the copper to be passed was fixed at 76, and an extra price being paid for copper of a still better electrical quality, the mean conductivity of the copper employed was actually 89.14. In many of the older submarine cables laid before this point had received attention, the conductivity was even as low as 30 to 40.

The wire thus prepared and built up was then covered by the Gutta-Percha Company with four separate coatings of selected gutta-percha, having alternate layers of Mr. Chatterton's viscid compound between each coating and next the wire, so as to insure thorough adhesion, and the total exclusion of air or any film of moisture between them. The weight of gutta-percha applied was 275 lbs. to the nautical mile.

115. To ensure the highest perfection in the joints of the various lengths of gutta-percha covered wire used, a plan of testing was adopted at the suggestion of Mr. Clark, which will be readily understood by reference to Figure 57. A gutta-percha trough, about 18 inches long and 1 foot deep, containing water, is insulated from

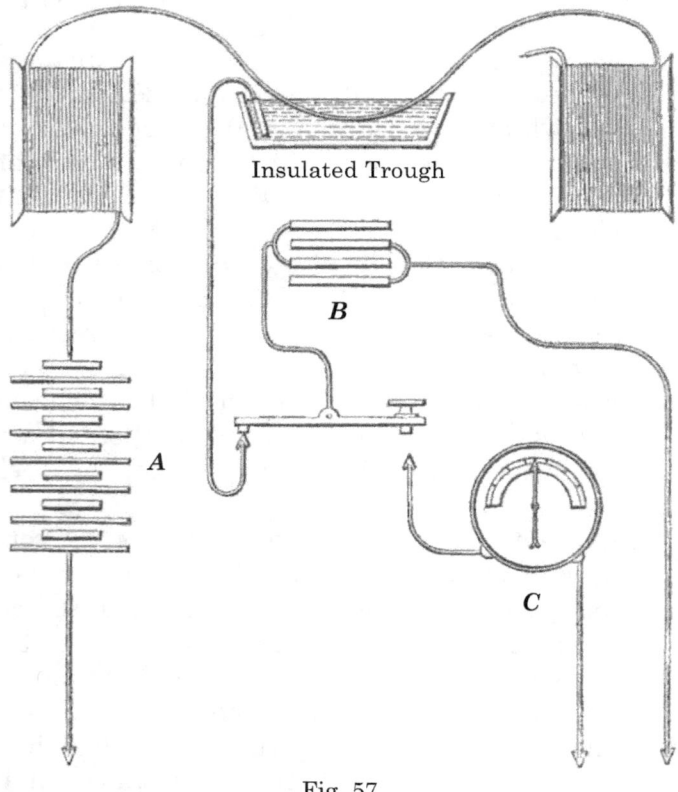

Fig. 57
Cable joint testing by accumulation

the ground by four legs of vulcanised india-rubber, or preferably, by suspending it from the ceiling by gutta-percha cords. The joint under examination is immersed in the water, and a battery of high tension *A* is connected to the copper conducting wire for five minutes, and all the electricity that escapes through the immersed joint during this time, passes into the insulated tank of water. To increase the capacity of the tank for storing up electricity, it is connected with an electrical condenser *B*, consisting of a great number of talc plates, coated on both sides with tinfoil, and having the same electro-static capacity as one mile of the cable. The escaping electricity accumulates in the condenser for the given time, and is then suddenly discharged through a delicate, suspended needle galvanometer *C*, and the number of degrees of the deflection furnishes an exact measure of the quantity of electricity passed through the insulator at the joint. In testing the joints of the Persian Gulf cable by this system, every joint was rejected and cut out whenever it gave less resistance than 40 feet of the core. Newly-made joints were found, almost invariably, to test perfectly; and it was only after, at least, twenty-four hours' immersion that a reliable test could be taken.

The joints made in the insulating material during manufacture, and in the finished core, have always been the subject of considerable anxiety to those engaged in the supervision of submarine telegraphs. It is essential for every atom of the wire submerged to be thoroughly protected, and although, the loss on a single joint may be so small as hardly to affect the tests obtained upon a considerable length, yet dearly-bought experience has shown, that the slightest defect may contain within it the seeds of a serious fault hereafter. The advantage of this system of testing will be apparent from the fact, that no less than thirty defective joints were rejected and replaced.

Throughout the manufacture of this cable the most searching tests were applied to the gutta-percha covering, and its wonderful excellence is shown from the fact, that the loss by leakage does not exceed one hundred millionth part of the current of electricity passing along the connecting wire in every nautical mile. This was a degree of insulation wholly unprecedented.

The outer sheath was put upon the core at Mr. Henley's factory, North Woolwich. A wet hemp serving was first wrapped round the gutta-percha wire, and 12 iron wires then laid over in the usual way, the hemp being moistened with the view of rendering any fault in the gutta-percha conspicuous to the continuous electrical tests maintained.

116. Two layers of yarn were then wound in opposite directions over the iron wires, being first saturated with a mixture of mineral pitch, Stockholm tar, and finely ground silica powder. This bituminous compound was applied, in a melted state; and while yet plastic, the outer coating was compressed by powerful grooved rollers, so as to form the whole into a smooth black rope combining flexibility with hardness, as shown in Figure 58.

While the outer iron armour was thus effectually protected from corrosion, the real destroyer of submarine cables, the hemp covering was at the same time preserved from tropical animalcules, by which it would be otherwise destroyed. Even the boring-tool of the teredo would recoil from the atoms of silica it would meet with upon any attempt at penetration.

The weight of the cable thus formed was nearly four tons per mile; but special shore ends of a more massive construction were also provided, weighing ten tons per mile. Sir Charles Bright undertook the personal control of the laying expedition, and the cable was conveyed to its destination by five sailing vessels of large tonnage, the *Assaye*, *Tweed*, *Marian Moore*, *Kirkham*, and *Cospatrick*, assisted by a steamer, the *Amber Witch*. — the latter being purchased for permanent service on the line,

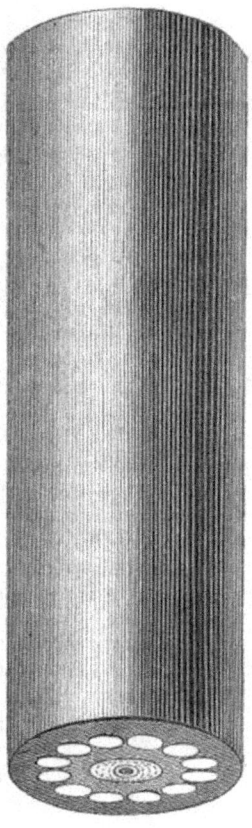

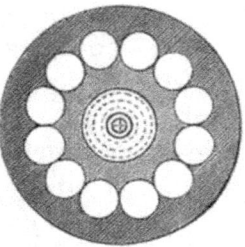

Fig. 58
Side view and section of the Persian Gulf cable

for repairing the cable in case of need, and for carrying stores and exchanging the staff between the stations. Each ship was provided with three iron tanks to hold the cable, which was kept constantly covered with water till passed into the Gulf.

117. The cable was safely laid in four sections: viz., Kurrachee to Gwadur, Gwadur to Mussendom, Mussendom to Bushire, Bushire to Fao.

The route selected in the Gulf, and across its entrance, had an exceedingly regular bottom; but on the Mekran coast, and especially between Kurrachee and Gwadur, it was by no means so good.

The submersion of the cable was commenced on the 4th of February, 1864, by Sir Charles Bright and his staff, Messrs. Laws, Webb, Woods, Alexander, and Mosley. Colonel Stewart led the expedition in the *Coromandel*. Nothing could exceed the perfect regularity with which the arrangements acted. The cable uncoiled itself with perfect freedom, and the laying of the different sections was completed on the 5th April. Her Majesty's steamers *Zenobia* and *Semiramis* successively took in tow the cable ships during the paying out; and the difficulties of communicating between the steamers and sailing vessels in tow, were overcome by signalling the ordinary Morse telegraph alphabet with a bull's-eye lamp; the shutter of which was fixed to a small lever, by which the duration of time the light was exposed could be varied at pleasure, and rapidly made to represent letters and words. By this plan messages were passed by flashes of light between the steamer and ship in tow, at the rate of twenty words per minute. The cable was paid out at the rate of from 5½ to 6 knots per hour, just sufficiently in excess of the rate of the ships to allow it to accommodate itself to the inequalities of the bottom.

Considerable difficulty occurred in landing the shore-end of the cable at Fao, and connecting it with the floating station moored off the entrance to the Tigris, owing to the shallowness of the water and extent of deep mud-banks. This part of the work has been thus described:—†

---

† *Times.* Account of the Indian Telegraph.

About five miles of cable, weighing some 20 tons, were distributed among ten of the largest boats belonging to the fleet. When about four miles had been payed out the boats grounded. Though there was very little water, there was a great depth of mud of about the consistency of cream. There was no use in hesitation, the cable must be landed at any risk; so Sir Charles Bright set an example to his staff and the men, and was the first to get out of the boat and stand up to his waist in the mud; an example which was followed by all the officers and men, upwards of a hundred in number, who were all soon wallowing in the soft yielding slush up to their chests, but still dragging the end of the cable with them. The progress through such a material was necessarily slow — half-swimming, half-wading; it was impossible to rest for a moment without hopelessly sinking below the surface, yet no one thought of abandoning the cable. Though it was only two o'clock when the party left the boats, it was nearly dark before the last reached the shore. All were grimed with mud, and nineteen out of twenty were nearly naked, having left or abandoned almost every article of clothing in the effort to reach the shore; but in spite of obstacles the cable had been landed. The troubles of the landing party were not yet over, for it was found that the ships of the expedition, which were waiting to receive them in the Tigris, were lying at the other side of a mud-bank, only a little less fluid than that which had just been passed, and four miles in extent. To make matters better, a thunderstorm, truly tropical in its violence, was raging; and the tide, which washes the banks, was rapidly rising. The party, however, made a dash for it, and all succeeded in reaching the ships, with the exception of one of the lascars, who was overwhelmed by the mud and tide, and sank before assistance could be afforded. The remainder were much exhausted, some few so much so, that they had to be carried by their companions.

It was expected that the Turkish line from Bagdad to Fao, at the head of the Persian Gulf, would have been completed simultaneously with the cable, but a considerable part of the broad tract of country, 400 miles in extent, intervening between the ancient city of the Caliphs and the miserable village of Shat-el-Arab at the junction between the Tigris and Euphrates, is inhabited by predatory tribes of Arabs, incessantly quarrelling one with another and mutually defying the Turks, their nominal masters. *Back-shish*, in the shape of subsidy or blackmail, was the only way to quiet these rapacious vagabonds; but it was not until the commencement of 1865 that their state of chronic revolt could be put an end to, and the line completed. In February, 1865, a telegram was received in London from Kurrachee in 8½ hours, and the communication was then opened to the public, and has since continued.

Although the connecting wires through Turkey, and between Kurrachee and the main Indian lines at Bombay are very badly managed, the cable has proved a commercial success, yielding, at present, over 100,000*l*. per annum upon its cost of about 400,000*l*., and there is no doubt this revenue would be very greatly increased if the connecting lines were better worked; for while the average time occupied in transmitting messages between Kurrachee and Fao over the four sections of the cable is only three hours, the messages between India and England frequently occupy many days in transit over the land lines. This is partly due to the utterly inefficient staff of half-castes employed in India by the Government; and partly to the carelessness and indolence of the Turks, who frequently allow their wires to be out of order for many days together; and of whose apathy we may judge, when it is stated that frequently messages have been so changed in their order of transmission, that those sent days after others have yet arrived first; probably from being filed as they arrived one after another at some intermediate station, and then, when sufficiently accumulated, sent on, taking those at the top first; thus utterly changing the order of transmission, and making the last message arrive first at its destination.

118. Endeavours have been made to arrange for through Turkish wires worked by English operators; but so jealous are the Sultan's Government of interference, that they will not sanction this necessary reform.

To obviate so serious an impediment to prompt and accurate communication, it is now proposed to carry a second line from England, by direct submarine cable to Gibraltar and Malta, to connect up the Malta and Alexandrian cable and Egyptian lines.

Thence by a cable to Aden and Bombay, so as to avoid the delays and errors arising from transmission at the hands of those working the present land route, comprising half-educated half-castes, Turks, Austrians, &c, who all combine in mutilating and mangling the plain English of our messages.

The subject was considered of such importance, that it was referred to a special committee of the House of Commons in 1866, consisting of Lord Stanley, Sir Charles Bright, Sir H. Rawlinson, Mr. Crawford, Mr. Ayrton, and others; by whom a report to the following effect was made:–

There exist at the present time two systems of telegraphy between England and our Indian possessions. One, commonly called the Turkish route, passes through Vienna and Wallachia or Servia to Constantinople, or reaches the latter city by way of Turin, across the Adriatic at Avlona, and along the shores of the sea of Marmora. Having reached Constantinople, it is forwarded through Asiatic Turkey to Bagdad, and thence to Fao, where it is received by British officers for despatch to Kurrachee, or by way of Teheran to Bushire, where it again falls in with the main line, and is then exclusively under the control of the Government of India. A message by this route may be dealt with by no less than ten administrations before passing into British hands. A large mass of evidence goes to prove that the vexatious delays constantly occurring on this line are mainly due to the Turkish officials, who not only neglect to keep any wire, or system of wires exclusively for Anglo-Indian use, under the terms of the Indo-Ottoman Convention, but fail egregiously in the prompt performance of their ordinary clerical duties.

The other available system, called the Russian route, passes through Berlin, via the Hague, thence into Russia through Tiflis to Julpha, on the Arras, and so over the Persian system to Bushire. These two systems of telegraphic communication are distinct as far as the head of the Persian Gulf, but inter-communication may be altogether suspended by the occurrence of any accident to the submarine cable between this point and Kurrachee. The Russian system appears to work well as far as the eastern frontier of that country, but fails in its entirety from the imperfect arrangements of the Persian service between Julpha and Teheran. A report submitted by Colonel Goldschmidt, in November, 1865, recommends that such part of the Anglo-Indian system as passes through the Ottoman Empire should be placed in the hands of Englishmen, making the principal stations at once Anglo-Ottoman, as at Fao. This plan, however, is inadmissible, for political reasons. The committee have therefore recommended that means should be adopted for facilitating the use of the Persian system with Europe, so as to bring the Russian route under efficient control; that to avoid future accidents, the Persian Gulf cable should be doubled, or, by arrangements with the Persian Government, that land wires should be instituted between Ispahan and Kurrachee. That the scheme of establishing a direct communication between Alexandria and Bombay, by way of Aden, is especially worthy of consideration, and should receive all possible aid from her Majesty's Government, as thus establishing a route under one management and responsibility. The committee also urge upon the Indian authorities the absolute necessity of improving their internal arrangements, so as to remove all risk of delay in the transmission of messages from Kurrachee to the interior.

119. The extension of the telegraph to Australia and China must speedily follow upon the communication between India and Europe. The telegraph is at present constructed as far east as Rangoon, from which there is no difficulty in selecting a cable route with a favourable bottom at a short distance from the coast to Singapore. Between Singapore and Hong Kong a cable can be readily carried in shallow water, touching at Saigon; or the connection with China may be effected by crossing the peninsula and laying a cable across the Gulf of Siam.

To effect the same object a land line of telegraph has also been warmly advocated by Captain Sprye, from Rangoon through Birmah and Western China; but in uncivilised countries communication by the aid of submarine cables, wherever practicable, is found far more reliable.

Proceeding southwards from Singapore towards Australia, the first section to Java can be laid in shallow water; and as a land line of telegraph has already been constructed by the Dutch Government throughout Java, a cable may be extended from the south-eastern extremity of that island to Timor, terminating at a station to be established at Coupang. With the exception of a short distance to the south of Timor, as yet not surveyed by soundings, the remaining link to Australia can be laid in shallow water.

The Australian telegraphs already extend between Adelaide, Melbourne, Sydney, Brisbane, and Port Denison, a distance of about 2400 miles, and are being pushed on northwards from the latter place towards the Gulf of Carpentaria; and as the whole of the intermediate country is being rapidly occupied by settlers, there will be little difficulty in completing the link between the Australian telegraph system and the landing point of the cable.

As a powerful company is in course of formation to carry out this vast scheme, it is probable that it may be realised within a few years; and when it is achieved, and our countrymen at the Antipodes are able to hold prompt correspondence with their mother country — but not till then, as Sir Charles Bright said at the recent Atlantic Cable Banquet at Liverpool, "may telegraphists triumphantly point to their works and exclaim:–

" *Quae regio in tenia nostri non plena laboria.* "

PLATE VIII

The *Great Eastern* picking up the Atlantic Telegraph cable of 1865

8
THE ATLANTIC CABLES
OF 1858, 1865 AND 1866

120. Projection of the line—121. Preliminary experiments on long underground wires.—122. Soundings of the Atlantic—123. The cable of 1857 and unsuccessful expedition of that year—124. The machinery employed—125. The successful expedition of 1858, and failure of cable a month after laid—126. Resuscitation of the enterprise, and arrangements for cable of 1865—127. Improved paying-out and picking-up machinery—128. The expedition of 1865; repeated faults in cable, and final fracture—129. The successful expedition of 1866, improved mode of testing, &c.—130. Atlantic signalling apparatus adopted by Professor Thomson—131. Congratulatory messages.

120. BY the end of the year 1855 the North American lines had been extended as far as Newfoundland, while in Europe the wires of the Magnetic Company extended to the extreme west of Ireland. The feasibility of uniting the two vast systems of European and American telegraphs by a submarine cable had engaged the consideration of several of the most enterprising telegraphists on both sides. It had been already proved that cables could be successfully laid in comparatively deep water, but the nearest points between the British Islands and Newfoundland are nearly 2000 miles apart; and the greatest length of submarine line submerged prior to 1856, would form but a fraction of such an enormous distance.

121. The most important question then undetermined was the possibility of working through such a length in an unbroken circuit, and at a speed that would enable messages to be passed in succession quickly enough to prove remunerative.

All doubts on this point were, however, removed by a series of experiments instituted by Sir Charles Bright, in conjunction with Mr. Whitehouse, upon the long lengths of underground gutta-percha covered wires belonging to the British and Irish Magnetic

Telegraph Company, which were so connected on various occasions as to afford a length of upwards of 2000 miles in one continuous circuit. Signals were clearly and satisfactorily transmitted this vast distance at the rate of 210, 241, and 270 per minute, with a facility that would answer every commercial requirement.

The difficulty in working that was found to arise from the retardation of the electric current by induction was overcome, as it had previously been in the magneto-electric instruments used by the Magnetic Company, by using a succession of opposite currents. By this means the latter or retarded portion of each current was blotted out by the opposite current immediately following it; and thus a series of electric waves could be made to traverse the wire one after the other, several being in the act of passing onward at different points along the conductor at the same time. Secondary currents were employed, generated by an induction coil, and the electricity was found to occupy about three seconds in passing from one end to the other.

The New York and Newfoundland Telegraph Company had previously obtained from the Governments of Newfoundland, Nova Scotia, Prince Edward's Island, and the State of Maine, the exclusive right to land cables, upon condition of their connecting those countries with the network of telegraphs already constructed in the United States. Mr. Cyrus Field, the Vice-President of that company, then entered into an arrangement with Sir C. Bright, Mr. Brett, and Mr. Whitehouse to combine in carrying out the undertaking.

122. The nature of the ocean's bed had become by this time well understood, several series of soundings having been taken by Lieutenant Maury and Lieutenant Berryman, of the United States Navy, and also by Commander Dayman, R.N., which proved that a gently undulating plateau of great breadth existed nearly the whole of the distance between Ireland and British North America, at a depth varying gradually from 1700 to 2300 fathoms. These depths, although very great, are insignificant when compared with the nearly unfathomable soundings further to the southward than the belt of ocean thus indicated, where the lead has been found to pass uninterruptedly downwards for 6000 and even 7000 fathoms.

The table-land thus raised at the bottom of the sea appears to follow the course of the Gulf Stream, and has been built up by the deposit of the shells of microscopic animalculae, which have rained down, as it were, for countless ages, carried along by the warm water of this "river in the ocean." The minute diatomaceae and globiferse die in myriads upon the temperature of the Gulf Stream being lowered by contact with the more northern seas; hence about Newfoundland, where the Gulf Stream is somewhat checked and first comes in contact with the cold waters from the north, a much more rapid deposit of these animalculse has taken place, resulting in the formation of the cod banks. When specimens of the soundings brought up are examined, they are found to be analogous to the material forming our chalk cliffs, which have also been built up in past ages by the same little creatures. The diatomacese in their action in the colder regions may, in fact, be likened to the coral insects of warmer climes, as regards the gradual building up of the sea into land. As these microscopic shells are so fragile that a breath would almost destroy them, they afford a proof that there are no currents moving over the surface of this plateau, for had the shells been rolled to and fro their delicate organism would have been bruised to pieces.

123. These facts promised permanent security to a cable when laid, and rendered it unnecessary to provide one of great weight or size, such as would have been requisite for any depths affected by currents. On the other hand the enormous depths in which the Atlantic line had to be laid, made it necessary to construct a form of cable that should be capable of sustaining five or six miles of its own weight in the water, when suspended vertically; so as to allow of laying to, if required, during submersion. At the same time the cable had to be heavy enough to draw itself freely from the hold, somewhat in excess of the ship's speed, and to sink readily so as to avoid the lashing of the waves in rough weather, and pass without interference through the currents near the surface. After experiments on upwards of sixty kinds of cables made by Messrs. Glass and Co., one was selected with a central conducting strand composed of seven copper wires, No. 22 gauge, and coated with three distinct layers of gutta-percha. This core was then surrounded with tarred yarn, and covered over with eighteen strands of iron wire as an outer protection, as shown in Figure 59.

The Governments on both sides having encouraged the project by guarantees of traffic, and the promise of vessels to assist in laying the cable; the promoters, in conjunction with the writer, made the experimental and other results public, by holding meetings at Liverpool, Manchester, and Glasgow. These led to the immediate formation of the Atlantic Telegraph Company; and nearly the whole of the capital, consisting of 350 shares of 1000*l.* each, was subscribed for in this country in the course of a few days, principally by those connected with the Magnetic Telegraph.

The engineering department was placed in the hands of Sir Charles (then Mr.) Bright, while Mr. Saward became secretary, and Mr. Whitehouse acted as electrician. The manufacture of the cable was at once proceeded with by the two well known contracting firms, Messrs. Glass, Elliot and Co., of Greenwich, and Messrs. Newall, of Birkenhead.

This, the first Atlantic cable, is shown in Figure 59. Lengths of ten miles for the shore ends were made exceedingly massive, weighing about ten tons to the mile, and encased with wires of great thickness. The British and United States Governments lent H. M. S. *Agamemnon,* and United States frigate *Niagara,* to carry the cable, with the frigates *Leopard* and *Susquehanna* to assist. The first attempt was made in August, 1857, with 2500 miles of cable on board. The expedition started from Valentia, Ireland, where the shore-end had been landed with great enthusiasm in the presence of Lord Carlisle, then Lord Lieutenant. After the paying out had gone on successfully

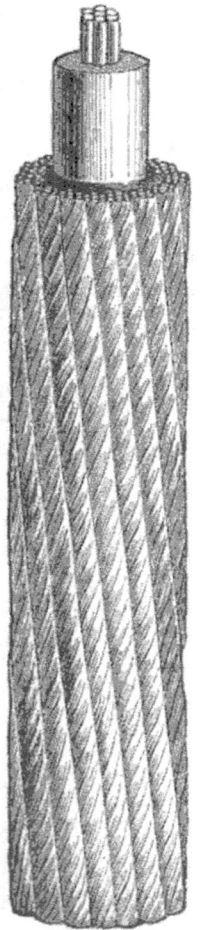
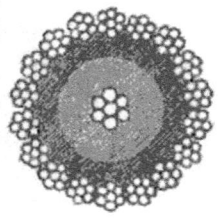

Fig. 59
Original Atlantic
cable of 1857-8

from the *Niagara* for several days, during which 380 miles had been laid, the cable parted at night during a strong breeze.

After the failure of this first attempt to span the Atlantic, many crude suggestions, to which the newspapers gave prominence, were made; principally by those unacquainted with submarine cable work. Amongst them may be noticed the general impression that neither a cable nor any other substance could sink to the bottom of the ocean; but that, after descending a certain depth, everything, even a cannon ball, would arrive at its point of flotation, owing to the increase of pressure and supposed increase of density of the water. In this exploded fallacy, the fact that water is less compressible than even iron, and therefore less capable of increase of density, was ignored.

It was suggested that the proper point to pay out a cable was from the centre of a ship, as the point of least motion, and therefore least liable to injure the cable, and it was proposed to have an opening in the middle to let it down; but as a cable in paying out leaves the ship at an angle but a little removed from the horizontal, the absurdity of such a proposition is manifest.

Another plan, strongly advocated, was to wind the cable upon an immense drum, to be towed by steamers while the line unwound. It never seems to have occurred to the suggestor that in rough weather, or even with a slight breeze, all control would be lost over such an unwieldy machine; while the strain upon the cable in unwinding it would be utterly beyond regulation.

A trail, or flexible pipe, was also proposed to hang from the ship's stern to the bottom of the sea, through which the cable was to be allowed to pass; but the promoters of this plan omitted to consider the effects of the friction resulting from 2000 miles of cable passing through it. Of whatever substance such a trail might be made, a day or two's rubbing of the cable would have worn it through.

124. In the next effort, of 1858, it was decided, upon the strong recommendation of Sir Charles Bright, that instead of commencing with the easiest and shallow part of the laying, from shore, that the ships should begin in mid-ocean, so as to start with the most difficult work. This had the double advantage of enabling them to select fine weather at the outset, and of reducing the time occupied in laying the line to one-half.

The paying-out machinery used in the first expedition consisted of a series of four consecutive grooved wheels, the cable passing from one to the other so as to embrace about two-thirds of the circumference of each. These wheels were geared together, and a friction-brake connected with the gearing, so that the person in charge of the machine could vary the check applied from time to time, as the surging of the ship drew out more or less cable. The machinery was considerably modified for the expedition of 1858, and is shown in Figure 60.

The cable on its way from the hold to the stern of the vessel was made to pass four times over two large drums, embracing half the circumference of each drum. Attached to the axles of each of the drums there were two wheels with friction straps, which exerted a given retarding power, according to the weights hung on to the levers tightening the straps.

Between the stern of the vessel and the machine the cable was bent somewhat out of the straight line, by being led under the grooved wheel of a dynamometer. This wheel had a weight attached to it, and could be moved up or down in an iron frame. If the strain upon the cable was small, the wheel would bend the cable downwards, and its index would show a low degree of pressure; but whenever the strain increased, the cable, in straightening itself, would at once lift the dynamometer wheel with the indicator attached to it, which showed the pressure in cwts. and tons. The principle is similar to the ordinary spring letter-weighing machine. The amount of strain with a given weight upon the wheel, was determined by experiments, and a hand-wheel in connection with the levers of the paying-out machine was placed immediately opposite the dynamometer; so that, directly the indicator showed strain increasing, the person in charge could at once, by turning the hand-wheel, lift up the weights that tightened the friction straps, and so let the cable run freely through the paying-out machine. Although, therefore, the strain could be *reduced* in a moment, it could not be *increased* by the man at the wheel.

This principle has been since used in connection with the oth-

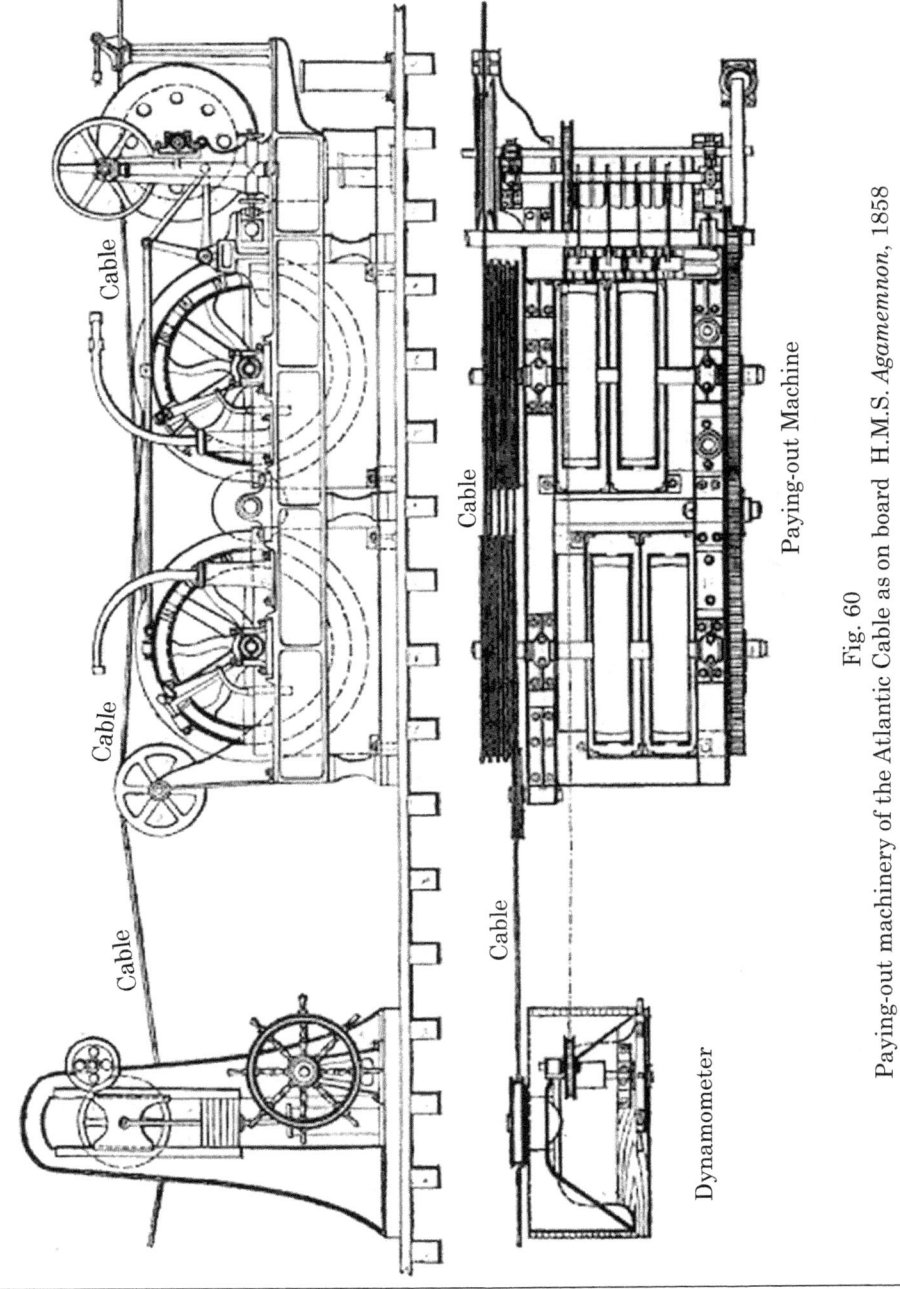

Cable

Cable

Cable

Cable

Cable

Paying-out Machine

Dynamometer

Fig. 60
Paying-out machinery of the Atlantic Cable as on board H.M.S. *Agamemnon*, 1858

er deep-water cable expeditions.

125. In June, 1858, the *Agamemnon* and *Niagara* proceeded to the rendezvous, accompanied by H.M.S. *Valorous* and *Gorgon*. They encountered a heavy gale, and the great weight of cable on board the *Agamemnon* nearly made her founder; but Captain Preedy's seamanship brought her through, and the squadron met at the rendezvous after sixteen days of danger and apprehension. Another attempt was then made; but after laying a small length the cable parted, and the ships returned to Queenstown. After re-coaling they started for another, and, this time, successful effort; the achievement of laying the cable between the two continents being completed on the 5th August, 1858, by Sir Charles Bright and the engineers forming his staff (Messrs. Canning, Woodhouse, Clifford, and Everett), after an uninterrupted and most arduous task of eight days, during which 5022 miles of cable were laid. The electrical arrangements during the laying of the cable were under the charge of Professor Thomson.

On the ends of the cable being landed at Valentia, Ireland, and Trinity Bay, Newfoundland, and handed over to the Company's electrician, Mr. Whitehouse, he found that signals passed from shore to shore with as great speed and strength as those transmitted through the folds of the cable before the expedition left England. After transmitting messages for nearly a month, some defect in the insulation of the conducting wire was found to interfere with the further working.

The testing showed a fault at a distance of about 270 miles from Valentia, the electrical leakage through which had been augmented by the strong currents used to pass signals through the cable.

During the period that the Atlantic cable of 1858 was in good order 366 messages, consisting of 3942 words, were interchanged through it between this country and America; 97 messages, of 1102 words, being forwarded from Valentia to Newfoundland; and 269 messages, of 2840 words, from Newfoundland to Valentia.

Among these may be instanced messages from Her Majesty to the President of the United States, and his reply; messages stopping the departure from Canada of two regiments for this country, thus saving at least £50,000 unnecessary expense to our Government; and messages announcing the safe arrival of the steamer *Europa*, with mails and passengers uninjured, after her collision with the *Arabia*.

The last words passed through were read at Valentia on the 20th October, 1858.

Several attempts were made to pick up the cable to the faults, but were unsuccessful; and this disastrous result, following so soon after the accomplishment of the great feat of connecting the two continents together, discouraged further enterprise in the same direction for a number of years.

The experience thus gained on many points was, however, of the highest importance. It was seen that with deep sea cables it was advisable to construct them proportionately stronger and specifically lighter than the first Atlantic line, so that they might be recoverable from great depths. It was also obvious that for so long an unbroken circuit the conductor should be larger and the gutta-percha insulation more perfect, so as to enable a greater speed of transmission to be attained with a less intense current. In fact, the weaker the electric charge capable of producing an effect at the other end, the less tendency it would have to burst its way through the gutta-percha at any defective point, and therefore the more likely the cable would be to last.

126. By the exertions and economical management of Mr. Saward, the Secretary, aided by the contributions of Mr. (now Sir C. M.) Lampson and other Directors, the Atlantic Telegraph Company was kept alive year after year, until the successes achieved with the Malta and Alexandria, the Barcelona and Port Mahon, and other deep sea cables, followed up by the connection of Europe and India through the Persian Gulf, gave a fresh stimulus to Atlantic enterprise; and, with the co-operation of Mr. (now Sir R. A.) Glass, capital was again forthcoming for the cable of 1865.

The form of cable chosen was very similar in construction to that recommended to Government by Sir Charles Bright, in 1859, for the proposed Falmouth and Gibraltar line; the principle being the combination of iron wire and hemp for the outer protecting strands, by which the specific gravity was reduced, and greater strength gained; while casing the wires in hemp, saturated with tar, would preserve them from rust. A section and side view of the cable of 1865 are shown at Figure 61.

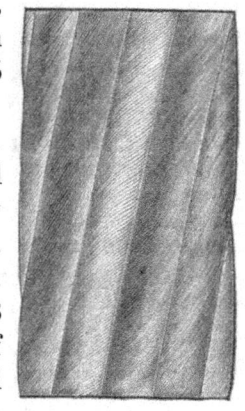

The weight of this cable in air was 35 cwt. 3 qrs., or about 15 cwt. more than the original Atlantic; while its weight in water was only 14 cwt., or but 1 cwt. more than the old cable. The copper conducting strand weighed 300 lbs. per mile, or nearly three times that of the 1858 strand, and was insulated by eight coatings of gutta-percha and Chatterton's compound laid alternately over one another. The former cable had but three coatings of gutta percha.

The rope of 1865 could withstand a strain of seven tons and a half, but the strength of its predecessor was only about half as great. The weight and bulk of the cable being so enormous when multiplied by the length to be made — 2300 miles — it was determined to take up the Great Eastern, which was then seeking employment almost in vain. By this arrangement the

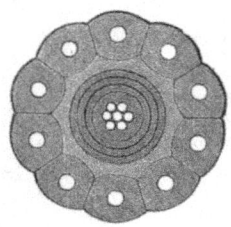

Fig. 61
Atlantic cable, 1865

whole of the cable could be stowed in one ship, while without her aid four ships of the largest size would have scarcely sufficed; and, as in 1858, the cable would have had to be much smaller in size. Even with present experience it would be a most dangerous experiment to attempt to lay a .cable piecemeal across the Atlantic from a series of vessels, as rough weather might at any time prevent the ends being successively joined as each ship finished its portion of the task.

The mission of this vast ship was at last discovered, and she was speedily prepared for the work. Huge tanks were built within her to receive the cable, and keep it continually saturated with water; so that in case the slightest fault occurred prior to the insulated cord passing into the sea, it would be at once detected.

127. The paying-out machinery was considerably improved by Messrs. Canning & Clifford, who had charge of the manufacture of the cable, under Mr. Glass' supervision, and of its subsequent submersion.

Messrs. Glass & Elliot having combined their works with those of the Gutta Percha Company under the title of the Telegraph Construction and Maintenance Company, the insulation of the conducting wire was proceeded with, *pari passu*, under the careful superintendence of Mr. Chatterton and Mr. Willoughby Smith. The eight separate insulating coatings reduced incalculably the chance of any defect occurring at one point in all, and resulted in the insulation being far superior to that of any previous cable.

The paying-out machinery fitted on board the Great Eastern was so arranged that, on the one hand, no strain could be put upon the cable in excess of that arranged beforehand, and adjusted by weights; but, on the other hand, the strain could be reduced or taken off altogether by turning hand-wheels similar to steering-gear. The principal hand-wheel was placed close to a dynamometer, or pressure-recording apparatus, similar to that used in 1858, (see Figure 60,) through which the cable passed on its way from the machine to the sea. If the index of the dynamometer showed any considerable increase of strain from any cause, the person in charge of it, and the paying-out machine, could at once take off the machine weights by turning the hand-wheel.

The paying-out machine is shown in Plate IX, but the mode of action will be more clearly shown by following the diagram, Figure 62. The cable, on coming up from the tank in the hold, passes along a conducting trough to the first of the six leading V wheels of the machine. It does not take a turn round this wheel **B**, but merely passes over the top of it and the five other wheels consecutively, being pressed down into their grooved rims by a small weighted wheel or jockey

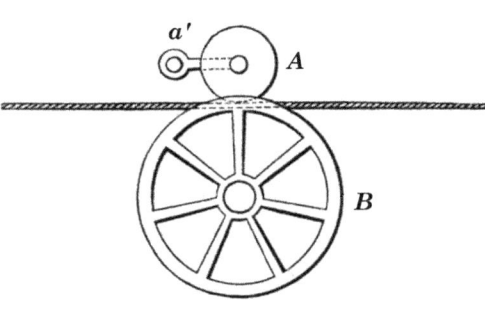

Fig. 62
Diagram of one of the six leading wheels
of machinery of 1865-66

pulley, **A**, around the circumference of which there is a band of india rubber, so as to produce a retarding effect upon the cable when necessary. The jockey pulley turns upon an axle at the end of an arm centered at **a'**, and the weights on the jockey pulleys can be released at once by turning a hand-wheel. After leaving the last of these wheels, the cable takes several turns round a large drum, the axle of which is connected to a brake arrangement similar to that of 1858, by means of which the speed of the drum with a given strain is checked or accelerated, according to the increase or reduction of a series of hand-weights, that can be attached or taken off as required.

Provision was also made for picking up the cable in case of accident. As ships will not steer stern foremost in operations of this nature, the head of the ship has to be kept to the cable, as it comes up from the sea. An auxiliary steam engine was fitted in the bows of the Great Eastern, geared to a pair of picking-up drums, round which several turns of the grapnel rope would be made. Another dynamometer (or pressure measurer) was placed between these drums and the bow sheave, to indicate the strain upon the grapnel rope, and thus show when the cable was hooked, or when the pressure was becoming so great during the hauling-in process as to imperil either the grapnel rope or the cable.

128. Thirty miles of very massive shore-end cable, weighing 20 tons per mile, having been previously laid by the steam-ship *Caroline*, from Foilhommerum Bay, Valentia, the *Great Eastern* spliced the deep sea cable to the gradually-tapered end of the shore cable, and commenced paying out on the 23rd July, 1865.

Early in the morning of the following day, when 75 miles from land, a fault was discovered in the insulation; and as the tests localised it to a few miles from the ship, the picking-up machine was soon set to work. After 10¼ miles had been drawn back the fault was arrived at, and the cause of injury found to be a piece of iron wire, about two inches long, and the same gauge as that covering the cable, which had by some means been driven into the gutta percha so as to touch the conductor. Complete leakage was thus produced after the cable had been sufficiently long under water to allow the moisture to penetrate the outer covering of tarred yarn.

Everything then went on well for the next five days, until the afternoon of the 29th July, when, after 716 miles had been laid and the ship was in soundings of more than two miles deep, a similar fault was discovered to have passed overboard. This was rectified in the same way as its predecessor by the tedious and difficult process of hauling back. These faults were referred by some to accident arising from the brittleness of the outer wire, probably at points where welded; but by others to malicious injury.

On the 2nd August, when at a distance of 1050 miles from Ireland, after 1186 miles of cable had been successfully laid, for the third time a loss of insulation was reported in soundings of 2¼ miles. The *Great Eastern* commenced drawing in the cable to get at the fault, which the tests showed to be about six miles off; but as the ship drifted, one of the projecting hawse-pipes at her bow chafed the cable, and before the injured part could be drawn through the machinery it broke. Repeated efforts were made to recover the lost cable with the picking-up machinery and grapnels; but the shackles of the grappling-rope were not strong enough, and gave way repeatedly after raising the cable some distance, in one instance nearly halfway up to the surface. Though the rope might be strong enough, the swivels connecting its length together were not; and the axiom that "nothing is stronger than its weakest part" received an apt illustration in this case. The battle with the ocean was over for the year: as length after length of grappling-rope had disappeared with the cable to the bottom, and left none on board for renewed attempts.

129. The experience derived from this expedition plainly showed that success with an Atlantic cable was not merely to be regarded as an accidental result, but might be really considered to be reduced to a certainty: and that with stronger tackle there was good reason to believe that the lost end of the cable might be recovered. Although most people unaccustomed to cable-work were incredulous, those connected with the undertaking were so sanguine of success that a new company, the Anglo-American, was organised to raise fresh capital and carry out the scheme in conjunction with the Atlantic Company.

As it had been found that the tarry composition coating the outside strands interfered
with the speedy detection of faults, by filling
up any injury to the gutta-percha and by preventing ready penetration of the water in
such cases; it was considered better to omit it
in the cable of 1866. With this exception the
new cable, which is shown in Figure 63, was
similar to its lost predecessor. Through this
composition being omitted, it weighed only 31
cwt., or nearly 5 cwt. less than the other.

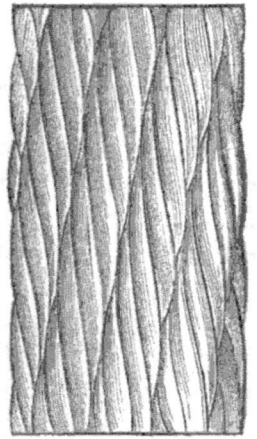

The tests for faults in the previous cable
were periodical — intervals being allowed
between each for passing messages between
ship and shore; but by an ingenious plan devised
by Mr. Willoughby Smith, it was arranged
that the connections of the cable of 1866
should be so made as to keep it under a continuous test for insulation, and yet allow
communications to continue between those
engaged at each end of the cable.

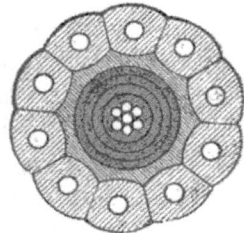

A diagram showing this testing arrangement is given in Figure 64.

Fig. 63
Atlantic cable, 1866

Suppose the leakage through the resistance R, connected
with the shore end of the cable, to equal that through the gutta-
percha of 4 miles of the cable, this amount of leakage would flow
through the galvanometer G' to earth, and thus a constant deflection
would be observed so long as the cable was kept charged at a uniform
tension. But should the tension be altered, either by the occurrence
of a fault in the cable, or for signalling, by reversing the current
on board ship or pressing down the key, K, on shore, a change in
the deflection of both galvanometers becomes at once observable.
By this means constant communication can be maintained with
shore, and any injury to the cable at once detected.

130. Let us now turn to the signalling department, and see
what arrangements were made for working the cables when laid.
In the apparatus applied by Mr. Whitehouse to pass electricity

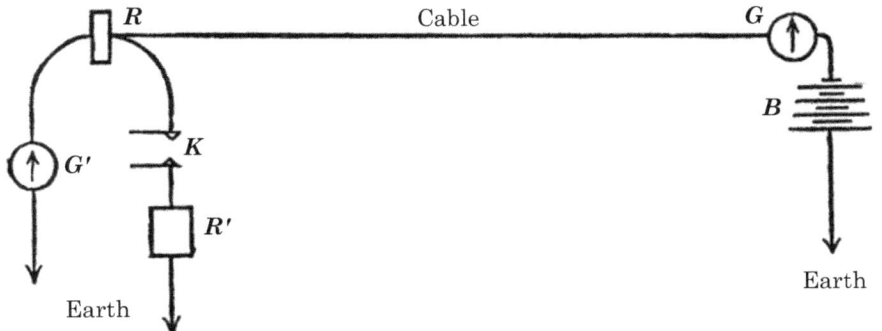

Fig. 64
Willoughby Smith's arrangement for continuous insulation test

B. Battery on board ship.
G. Thomson's galvanometer, through which the battery is connected to the cable.
G', Another galvanometer on land in connection with resistance **R**.
K. A contact key by which the current from the cable can be turned through a comparatively low resistance **R'**.

through the cable of 1858, large induction coils were used, and the secondary current produced was probably equal in power to that derived from 2000 cells of an ordinary Daniells' battery. The comparatively low state of insulation of that cable could not long withstand this excessively intense current, and the cable gradually ceased transmitting signals. The electricity forced its way through incipient points of leakage in the gutta percha envelope, and converted them into fatal faults. Such an untoward result taught electricians the value, or rather the necessity, of moderating the power used in working such a long circuit. With the view of facilitating this, as we have seen, a much larger conducting wire, and a more perfect gutta-percha envelope were adopted. Taking advantage of these improvements in the conductor, Professor (now Sir William) Thomson adopted a very ingenious means of producing a full and visible signal from an extremely minute movement of a magnetic needle; and was thus able to turn to account the feeblest electricity arriving through the cables.

The apparatus, which is similar in principle to Gauss and Weber's telegraph of 1837, consists of a small and exceedingly light steel magnet with a tiny reflector or mirror fixed to it— both together weighing but a single grain, or thereabouts. This delicate magnet is suspended from its centre by a filament of silk, and surrounded by a coil of the thinnest copper wire, silk covered.

When electricity passes through this surrounding coil of wire, the magnet and mirror take up a position of equilibrium between the elastic force of the silk and the deflecting force of the current from the cable circulating through the coil. A very weak current is sufficient to produce a slight, though nearly imperceptible, movement of the suspended magnet. A fine ray of light from a shaded lamp behind a screen at a distance is directed through the open centre of the coil upon the mirror, and reflected back to a graduated scale upon the side of the lamp-screen turned towards the coil. An exceedingly slight angle of motion of the magnet is thus made to magnify the movement of the spot of light upon the scale, and to render it so considerable as to be readily noted by the eye of the operator. The ray is brought to a focus by passing through a lens. By combinations of these movements of the speck of light (in length and duration) upon the index, an alphabet is readily formed.

The magnet is brought back to zero after each signal by the magnetic action of the earth, or else by the use of a small adjusting magnet.

The plan usually adopted for reinforcing the effect of a current on ordinary lines of telegraph is to let the magnet deflected (or soft iron attracted) make contact with a metallic stud, and thus bring into play a local battery to produce a more marked signal. With the 2000 miles circuit of the Atlantic cable, however, it was desirable to use currents of such small power, that the signal produced would not suffice for the firm contact requisite to turn on the local battery. The introduction of the mirror system rendered this unnecessary, through multiplying and magnifying the Atlantic signal by the agency of imponderable light!

This plan was put in practice with the Atlantic cable of 1858; and the messages then transmitted were read by the receiving clerk holding down the key of a Bain's recording instrument, whenever the ray of light began to move from zero upon the scale; as soon as it commenced returning to zero the clerk released the key. Thus marks and blanks were produced by decomposition up-

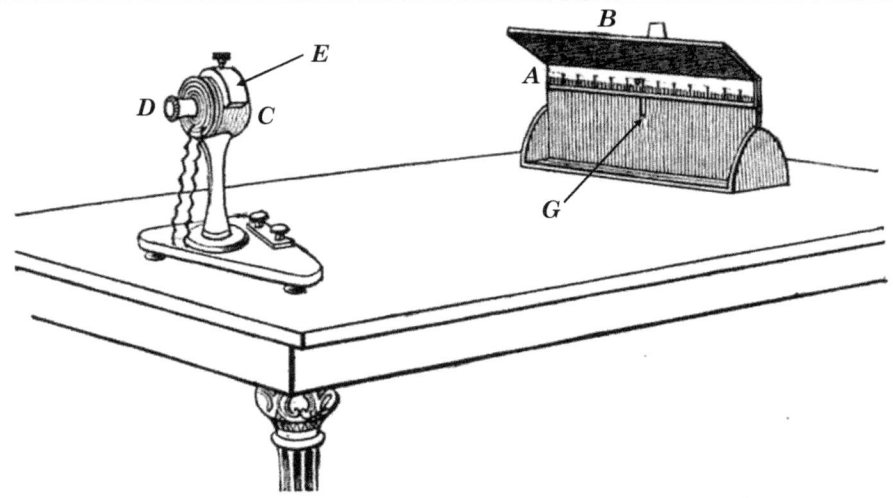

Fig. 65
Atlantic telegraph, the reflecting galvanometer and screen

on the chemically-prepared riband of the Bain's apparatus, corresponding with the movement of the light, and letters were formed by these combinations of conventional marks. The clerk therefore took the place of a relay, in making a permanent record of the duration of each movement of the light. The arrangement of Professor Thomson's apparatus is shown in Figures 65, 66 and 67.

Figure 65 shows the galvanometer **C** mounted upon a stand with the screen **B** and its graduated scale **A** at a short distance. A lamp is placed at the back of the scale; and immediately below the zero mark of the scale a small slot **G** is cut, to allow a streak of light to pass from the lamp to the small suspended mirror in the galvanometer coil; **D** is a thumbscrew to regulate the position of the mirror magnet, and **E** is a small adjusting magnet outside the coil.

Figure 66 shows the galvanometer coil **F** with its suspended magnet, **G** and mirror **H**.

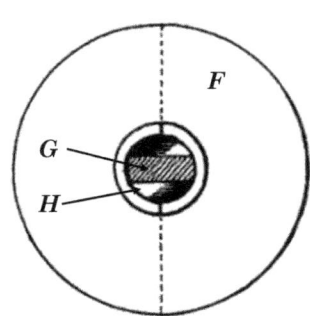

Fig. 66
Atlantic galvanometer coil, mirror and magnetic needle

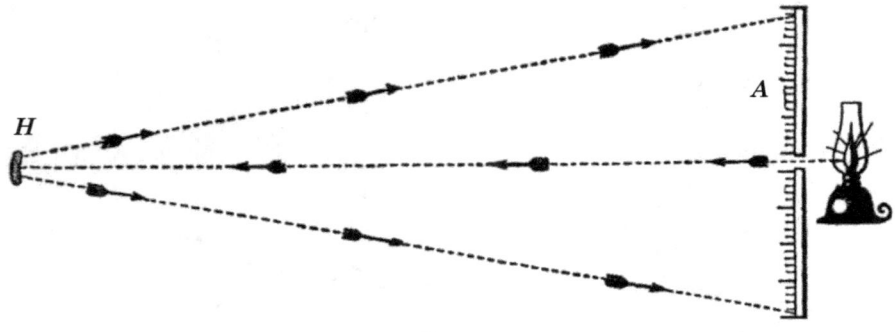

Fig. 67
Path of reflected light traversing graduated scale

Figure 67 is a diagram showing the manner in which the slightest movement of the mirror **H** makes the ray of light reflected from it traverse the graduated scale **A** at a distance.

When this apparatus is used on board ship, local magnets are employed to keep the mirror in the zero position; and the needle and mirror being carefully adjusted, with the centre of gravity in line with the silk filament to which they are attached, the signals keep steady even when the vessel is rolling heavily.

All the arrangements for the cable of 1866 being complete, on the 13th July the *Great Eastern* completed the splice of the main cable to the 30 miles of shore end off Valentia, and at 3.20 p.m. the Atlantic cable began to pass over the V wheel at her stern.

The laying of the cable, as in 1865, was entrusted to Mr. (now Sir Samuel) Canning and Mr. Clifford, assisted by Mr. Temple. The navigation of the great ship was in charge of Capt. (now Sir James) Anderson, assisted by Staff-Commander Moriarty, R.N., and Mr. Halpin; and the electrical department was represented by Professor (now Sir William) Thomson and Mr. Willoughby Smith, the chief electrician of the Telegraph Construction Company. Mr. Cyrus Field, Capt. Hamilton, R.N., and Mr. (now Sir D.) Gooch, M.P., directors of the Anglo-American Telegraph Company, with Mr. Deane, their secretary, were also on board. In addition to the cable required for the line of 1866, a sufficient length was taken to complete that of 1865, should the end be recovered as was confidently believed by those in charge.

The great ship was accompanied by H.M.S. *Terrible* and the steamers *Albany* and *Medway*, the two latter having picking-up apparatus on board to assist in fishing for the lost cable.

The expedition was favoured by fine weather, and everything went smoothly till the 18th July, when the cable became entangled in the hold through one flake fouling another. The paying-out part of the coil caught three turns of the cable immediately under it, and drew the bights into the eye of the coil in a confused tangle. The *Great Eastern* was fortunately brought up in time to prevent the huge knot of cable entering the machinery, and in the course of a few hours the confused mass was unravelled, and the work proceeded. Messrs. Canning and Clifford were well acquainted with foul flakes before, and knew how to set matters straight again.

No further interruption occurred, and the American end of the cable was successfully landed at Heart's Content Bay, Newfoundland, on the 27th July, in perfect order. Messages conveying the glad tidings were at once sent to Mr. (now Sir B. A.) Glass, the managing director of the Telegraph Construction Company, who had patiently and anxiously watched the progress of this great work — the carrying out of which was in a great measure due to his untiring energy and determination.

131. The following are some of the congratulatory messages passed.

As soon as the telegraphic communication between this country and America was completed, a message was transmitted through the Atlantic cable from her Majesty to the President of the United States:—

> From the Queen, Osborne, to the President of the United States, Washington.
> July 28th, 1866.

> The Queen congratulates the President on the successful completion of an undertaking which she hopes may serve as an additional bond of union between the United States and England.

The President replied as follows:—

From Andrew Johnson, the Executive Mansion, Washington, to her Majesty the Queen of the United Kingdom of Great Britain and Ireland.
July 30, 11.30 a.m.

The President of the United States acknowledges with profound gratification the receipt of her Majesty's despatch; and cordially reciprocates the hope that the cable, that now unites the eastern and western hemispheres, may serve to strengthen and perpetuate peace and amity between the Government of England and the Republic of the United States.

The following telegram was forwarded by the Earl of Carnarvon to Viscount Monck, Ottawa, Canada:–

I am commanded by the Queen to convey to the Governor-General of her North American provinces her Majesty's congratulations on the completion of the Atlantic telegraph, and the strengthening thereby of the unity of the British empire.

Her Majesty includes her ancient colony of Newfoundland in these congratulations to all her faithful subjects.
July 28, 1866.
Carnarvon.

The answer returned by the Governor-General of Canada was as follows:–

Viscount Monck to the Earl of Carnarvon.
Cable. — Your message of July 28 received. Present my humble duty to the Queen; and assure her Majesty, that her Majesty's gratification at the additional strength which the completion of the Atlantic telegraph will give to the unity of her empire, is shared by all her subjects in British North America.
Ottawa, August 1.
Monck.

The Lord Mayor of London telegraphed to the Mayor of New York:–

May our commerce flourish, and may peace and prosperity unite us.

The Mayor of New York replied:–

The energy and genius of man, directed by the providence of God, have united the continents. May this be instrumental in securing the happiness of all nations and the rights of all people.

An exchange of courtesies also passed between the Mayor of Vancouver's Island and the Lord Mayor, by telegraph; the one trying to embody in his message the feeling of the colony, and the other that of the mother-country. The first message was received at the Mansion House, London, on the 3rd August, and was in these terms:—

> Franklyn, Mayor, Vancouver, July 81, to Lord Mayor, London.
> The infant colony Vancouver, 8000 miles distant, sends telegraphic cordial greetings to Mother England.

To this the Lord Mayor of London, taking up the vein, and reciprocating the sentiments of his far distant correspondent, replied by telegraph as follows:—

> To the Mayor of Vancouver's Island.
> Mother England acknowledges the cordial greeting of her infant son Vancouver. May peace, good will, and unanimity unite and prosper our happy family.

It will be observed that three days were occupied in the transmission of the message between Vancouver's Island and this country. It would be carried across the continent of America to Newfoundland — how far by telegraph does not appear. Seven hours, or thereabouts, would be spent in conveying it by steamer from Cape Breton Island to Newfoundland (between which the local cable was out of order), a distance of 70 miles; and thence by the Atlantic cable to Valentia, and on to London about an hour more, making three days and nights, or 72 hours. Altogether it must have travelled, taking the whole route, at the rate of upwards of 111 miles an hour, but in the 70 miles from the American shore to Newfoundland it would only be conveyed at the rate of 10 miles an hour.

Mr. Latimer Clark, C.E., the engineer of the Atlantic Telegraph Company, tested the cable on behalf of the Company, and forwarded the following report to their secretary.

Valentia, July 30, 1866.

My dear Sir, — Herewith I have the pleasure to forward you a certificate of the completion of the 1866 cable, which is similar in every respect to the one which, after consultation, has been given to the Anglo- American Telegraph Company.

The cable, as you have doubtless already heard, is in a most satisfactory electrical condition, and with the apparatus now used gives fully six words per minute. With improved apparatus it will give much more, and by the use of codes I expect fully thrice that speed will be obtained. The present signals are strong and excellent.

Before leaving Sheerness the insulation of the cable was 713 millions Siemens units per knot. This steadily increased as the cable was paid out, and is now as high as 2, 300 million units per knot. This increase is due partly to temperature and partly to pressure, and I will shortly calculate the influence of these combined causes, and give you the insulation of this cable and the Persian Gulf cable at a uniform temperature and pressure for comparison.

It is to be regretted that the Newfoundland line is not in an efficient state, as that would have added greatly to the *éclat* of the opening. We have one and all formed the most sanguine opinions as to the permanence and security of the line, as well as concerning the probability of repairing the 1865 cable, which we quite expect to see put to work again shortly.

It is almost unnecessary for me to congratulate you on the success of this part of the work, and the wonderful prospects of pecuniary success which the first day's work has opened out to us.

> Believe me, ever yours truly,
> Latimer Clark.

Mr. G. Saward, Secretary, Atlantic Telegraph Company.

PLATE IX

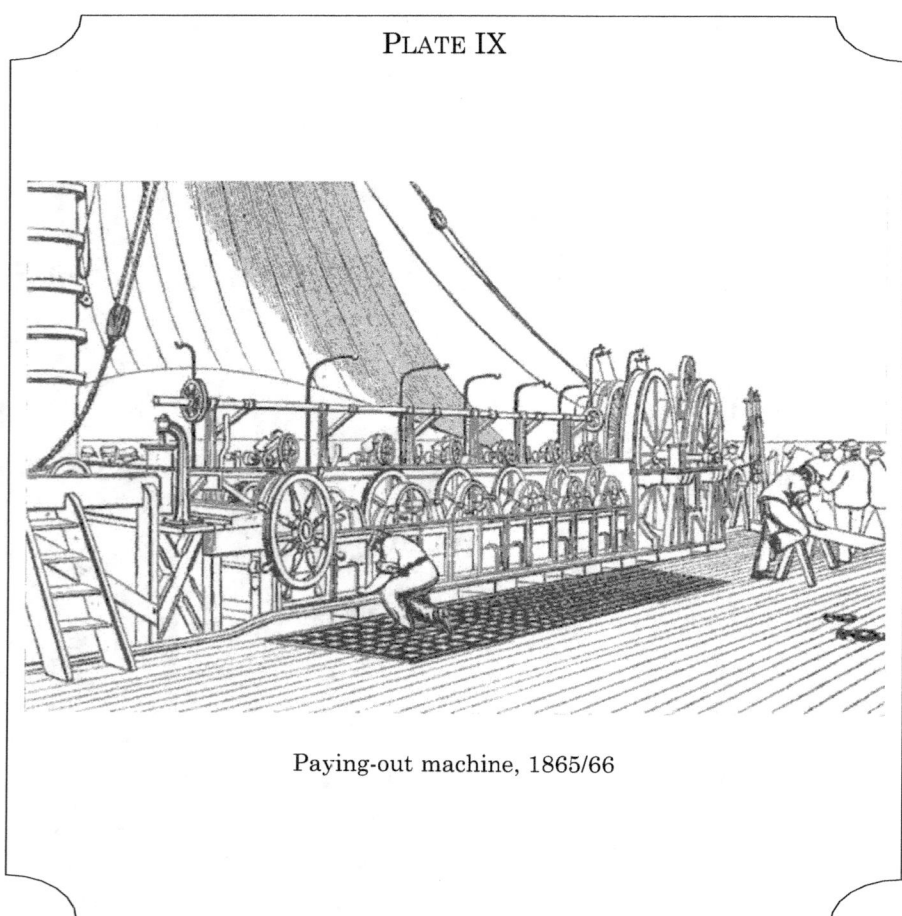

Paying-out machine, 1865/66

9

THE ATLANTIC CABLES
RECOVERY OF 1865 LOST CABLE

131. The practical difficulty of raising the cable—132. Appliances employed—133. The search; repeated efforts to recover it; and final success—134. Condition of the picked up cable and its perfect insulation—135. A lady's thimble used by Mr. L. Clark to form a battery to signal through the cables—136. Compensating keys—137. Tariff for Atlantic messages—138. Honours conferred on those engaged in the great work—139. Experiments of the Astronomer Royal and Dr. Gould to determine American longitudes by the cables—140. Electric current produced in a percussion cap passed through the cable.

131. THE Atlantic squadron then proceeded to search for the cable of 1865; and now began their most arduous task. Failure had been confidently predicted by many able men, some of whom went so far as to describe their grounds for considering the attempt impracticable.

The difficulty of the undertaking may be readily conceived, when we consider that a submarine cable when laid forms nearly a straight line upon the bottom; but in raising a bight of it to the surface, a considerable length beyond that upon the bottom is required to form the two curved sides, subtending the angle brought up. This will be more clearly seen from the following diagram, (Figure 68, p. 158), where *K* shows the line of a cable on the bottom of the sea, forming the base of the triangle, and *L*, *L'*, the two curved sides formed in bringing the cable upward. In laying deep-sea cables, 10 to 15 per cent of slack is usually paid out, and this surplusage would of course assist in forming the two sides of the bight. In such depths as the Atlantic, this excess would not however be sufficient; and it was therefore arranged to lift the cable partially at several points near to one another at the same time, by the co-operation of the three ships. If this could be accomplished, and the outermost vessel then broke the cable by

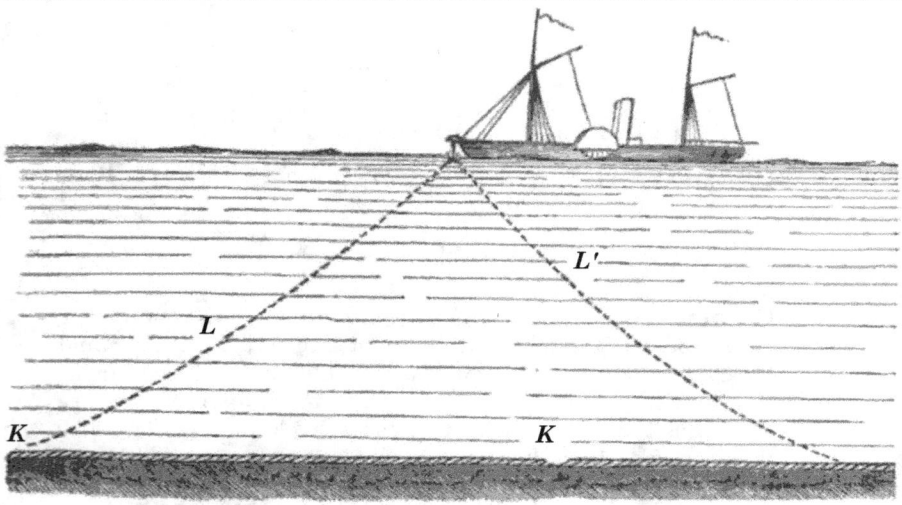

Fig. 68
Diagram showing by the dotted lines the angle a deep-sea cable has to make to come up to the surface

putting on additional strain, a length would be left free to form the outer side of the bight, and the cable would come up readily.

132. Prior to entering upon our history of the angling for such a great prize, let us look at the arrangement of tackle which was destined to recover from ocean depths of more than two miles the value of half a million sterling.

The line devised for this wonderful bottom fishing consisted of a combination of steel wire and hemp strands spun together. The grappling rope complete, measured 2½ inches in diameter, and was built up of seven smaller ropes (six laid round one), each composed of seven wires served with tarred hemp. The rope complete, therefore, consisted of no less than 49 wires, each isolated from its neighbour by the yarn covering it. The aggregate strength of this bundle of steel fibres was sufficient to bear a strain of no less than 30 tons. The hemp reduced greatly the specific gravity of this huge rope in water, while giving increased strength and elasticity; so that though weighing about 8 tons per mile in air, it was but 3 tons when submerged, and thus only put a strain of about 8 tons upon the picking-up machine when 2½ miles with the grapnel attached were hanging down in the ocean.

For fish-hooks, a number of five-pronged grapnels of the shape shown in Figure 69 were on board, weighing from 2½ to 4 cwt. each. Projecting springs were so attached to the grapnel shank, as to prevent the cable leaping up when once secured in the tenacious grasp of the flukes. The picking-up machine consisted of a pair of large drums at the bow of the vessel, as already described, geared to a powerful "donkey" engine, by which they could be made to revolve in either direction as required, like the winch of a fishing-rod.

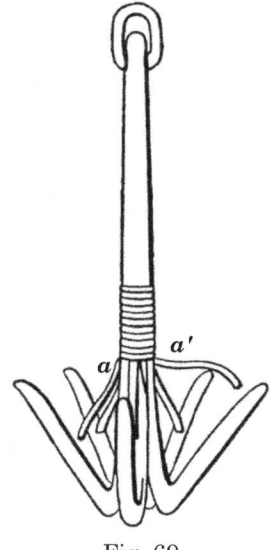

Fig. 69
Atlantic grapnel
a, a', are springs to prevent the cable surging from the grapnel flukes when hooked

To complete the similitude, between these drums and the bowsheave the grappling-rope passed under the wheel of a dynamometer, — the duty of which was precisely analogous to that of a fishing-float, to give warning of any nibble. A bite in this case was indicated by a tug on the line to the extent of an additional 3 tons, when the bight of the cable below had been hooked. The strain then ran up from the 7½ or 8 tons, due to the pendant grappling-rope, to 10½ or 11 tons when the prize was caught. Our readers can fancy the excitement on board upon a nibble being shown by a bob of the dynamometer index!

133. The questions now to be solved were:– Could the lost cable be found; and if found, could it be secured. The observations taken principally by Capt. Moriarty last year formed the sole clue as to the point in mid-ocean where the cable slumbered. The skill of Capt. Moriarty and Capt. Anderson soon set the first doubt at rest, by unerringly guiding the expedition to the spot where it had been lost. The *Albany* grappling-ship, with H.M.S. *Terrible*, made their way first to the rendezvous, in longitude 38.50 west, and commenced the search for the tiny rope in a depth of 14,000 feet of water, or nearly the height of the peak of Mont Blanc. The *Albany* soon hooked the cable, and on the 10th August lifted it some distance and attached a buoy. In the night, however, while a heavy sea was running, the buoy-chain parted, and the cable went to the bottom again.

This was the commencement of a most exciting hunt. On the 12th, the *Great Eastern* and *Medway* arrived; the great ship drew up the rope more than half a mile on the 15th, but in the act of buoying the rope slipped. Two days after she again got hold of the cable, and this time raised the bight above the surface to the bow-sheave. A hearty cheer greeted its appearance, but had scarcely died away when the cable was once more lost; the weather being too rough for the boats to co-operate in securing it, the cable parted before it could be brought in-board. Between the 17th and the 27th August, the different ships of the squadron repeatedly grappled it, the *Albany* raising it again to the surface on the 26th; but through boisterous weather they failed to secure the prize. In one instance, when the cable broke away, a man was caught by the grapnel-rope flying back, and hurled many feet from the fore castle-framing down to the deck below.

These repeated failures left them, as Capt. Anderson afterwards remarked, not only "shattered in ropes but nearly shattered in hopes."

As the cable at the bottom, where they had been so long working in the neighbourhood of longitude 38.40 west, was by this time greatly fouled and encumbered with the various grapnels and ropes which had given way in the many efforts to raise it; and as the depth of water was somewhat less at the point where the previous day's observations had been taken by Capt. Moriarty during the expedition of 1865, it was resolved to proceed to that point and try again. The exact spot was again indicated by his great nautical skill, and on the 1st Sept., operations were recommenced at longitude. 36.70 west, in about 11,000 feet of water, and fortunately in calm weather. The cable was soon caught by the *Great Eastern*, lifted 1¼ mile from the bottom and buoyed. She then shifted ground a few miles to the westward and at night again hooked it. The *Medway*, at the same time grappled the cable 2 miles further west, and was signalled by flashes of light to haul up quickly, so as to break it, and thus take the strain off the portion the great ship had hold of: she did so, and the bight then came in

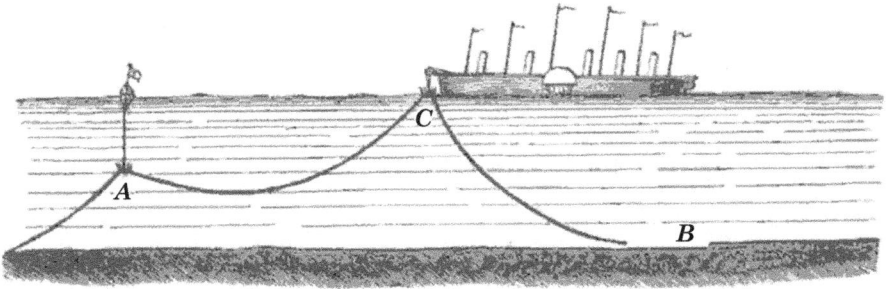

Fig. 70
Diagram showing the final arrangement for picking up the cable of 1865
A, Point where cable was buoyed to east by *Great Eastern*
B, Point where cable was broken by *Medway*
C, Bight of cable brought to the surface

readily but slowly, as if reluctant to leave the soft ocean bed upon which it had been so long reposing (Figure 70). The vast ship hung lightly over the grappling-rope, as if fearful of breaking the slender cord which was clenched in the iron clasp of the grapnel flukes 10,000 feet down in the depths of the sea. With a strain of 11 tons upon it, the tough unyielding fishing-line came in over the bows as rigid as a bar of iron; and as "slow but sure," is an axiom in cable fishing, so slowly but surely coil after coil of the huge grappling rope was drawn on board by the picking-up machine; until at last, amid breathless silence, the long lost cable, for the third time, made its appearance above the water. The voices of Mr. Canning, Mr. Clifford, and Capt. Anderson were alone heard, as their arrangements were made to put huge hempen stoppers over the cable, which was speedily attached to a five-inch rope, and having been released from the tenacious grasp of the grapnel, was hauled in by the machine after cutting away the western end of the bight.

The precious cable end was carefully led into the instrument room, where Mr. Willoughby Smith applied the tests in the presence of Mr. Cyrus Field, Mr. Canning, Mr. Clifford, Capt. Anderson, Prof. Thomson, Capt. Hamilton, Mr. Gooch, Mr. Deane, and others. In a few minutes their suspense was relieved by their tests showing the cable in good order, and immediately afterwards the answering signals arrived from the Telegraph Office at Valentia, and were received with loud cheers that re-echoed throughout the ship.

Let us now look at those patiently watching day after day, night after night, in the wooden telegraph cabin on shore. Such a length of time had elapsed since the picking-up expedition left Newfoundland, that the staff at Foilhommerum were almost hoping against hope. Suddenly, on Sunday morning the 2nd Sept., at a quarter to six, while the tiny ray of light from the reflecting galvanometer was being watched, the operator observed it move to and fro upon the scale. A few minutes afterwards the unsteady flickering was changed to coherency; the long speechless cable began to talk; and the joyful assurance arrived: "Canning to Glass, Valentia. I have much pleasure in speaking to you through the 1865 cable. Just going to make splice."

The glad tidings were also sent from the ship via Valentia to London, Newfoundland, and New York; so that for the first time it happened that men, tossed about on a stormy sea in utter darkness, could hold a conversation at the same time with Europe and America.

134. The cable when brought up was parti-coloured like a snake, half grey with the ooze of microscopic shells on which it had rested, and half black; showing that it had not thoroughly sunk into the material forming the bottom of the Atlantic, but had rested undisturbed and only half covered.

After splicing the end to the spare cable on board, the rest was laid successfully without hitch or difficulty to Newfoundland, on the 8th of September; forming the second line of communication with America. This cable tested upon completion even better than that of 1866; owing to the gutta percha of the 1200 miles laid in 1865 having become gradually consolidated by the continued pressure of the enormous weight of water, and to the uniformly low temperature (about 39° Fahr.) of the bottom of the sea in those great depths. The cable of 1866 has also steadily improved in insulation since it was laid, until now both cables test nearly equal.

Nothing, in fact, could be more favourable to the longevity of a submarine cable than these conditions, coupled with the level shell-strewn bottom and utter absence of motion in the lower waters of the Atlantic; and as the first cables laid nearly sixteen years ago have been found to last to the present time, with the gutta percha covering their wires as good and fresh as when laid, there appears no valid reason why any limit should be put upon the longevity of the two American cables, laid as they are utterly out of harm's way. In addition to this, the wonderful result of the last expedition has proved that cables, if injured, can be recovered from the greatest ocean depths; giving an assurance of permanent working — since repairs are practicable — and hence of a permanent value which such lines did not previously possess. Before this successful operation the grave doubt always arose: "But if it breaks, is not the injury irremediable?" Now, however, electrical science points out the exact distance of any fault from land, and as surely engineering skill can be brought to bear to set it right again.

135. To show how thoroughly perfect their insulation is, Mr. Latimer Clark had the extremities of the two conducting wires which now stretch across the Atlantic joined together in Newfoundland, so as to form an immense unbroken loop-line of 3700 miles. He then put some acid in a lady's silver thimble with a small piece of zinc and another of copper; and by this simple agency he actually succeeded in passing signals through the entire length of both cables in little more than a second of time. Of course the success of an experiment like this was possible only with a conductor as large and as wonderfully perfect in insulation as that of the Atlantic cables. The feat, however, forms a strange contrast to the enormous electrical power used in working the cable of 1858, when at first the intense secondary currents derived from the inductive action of 50 cells of a very large battery were employed; and afterwards a power equal to 500 cells, producing a current almost akin in its effects to lightning! There is no doubt that with the comparatively small conductor and poor insulation of the 1858 cable, an usually high power was requisite to drive the signals through in tolerably quick succession to form messages; but this energetic force soon wrought destruction to the very channel through which it passed, much as its prototype lightning blasts and destroys the conducting fibres of the tree by which it is conveyed to the earth.

136. In the practical working of the present Atlantic cables only 20 cells are usually employed, and as it was considered necessary to clear the cable of each signal promptly to allow the next to go through, various forms of compensating keys were devised by Sir Charles Bright, Sir William Thomson, Mr. Whitehouse, Mr. Varley, and the writer. It is found in practice that, instead of first sending a positive current of sufficient duration to produce the signal, and then blotting out the residual electricity it leaves behind in the cable by an equal negative current, it is better to transmit in rapid succession a series of opposite currents of very small duration. By this plan the cable is cleared much more rapidly, and consequently an increased speed attained for messages.

Figure 71 shows the arrangement of one of these keys, which is technically known as the "curb key," from curbing the successive currents. The apparatus consists of the key **N**, which when at rest locks the cam **B** by means of the detent **G**. On the key being pressed down it lifts the detent and allows the cam, which is attached to an axle driven by clockwork, to revolve. As the cam rotates projections from it successively raise and release contact springs, putting the battery poles in circuit with the cable and earth alternately in opposite directions, and each contact is thus maintained for the exact time necessary to produce the calculated

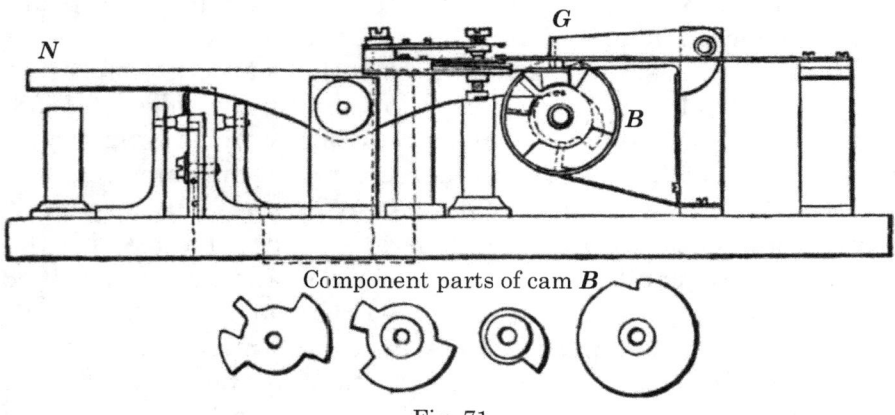

Component parts of cam **B**

Fig. 71
Transmitting apparatus proposed for the Atlantic Cable

effect. The number of projections in the cams and the arc subtended by each are of course determined by experiment, and are varied with the nature of the signal to be produced. If, for instance, a dot and dash alphabet similar to the Morse recording system is used, two keys are employed, one releasing a cam to make a dot by means of compensating currents of short duration, say 5 per cent; and the other a cam arranged to transmit the long mark, or dash, by contacts of longer duration, say 20 per cent.

The strength of the signals may also be varied by applying a greater or less battery power, and thus four distinct effects may be shown at the distant end of the cable: for instance, (1) a feeble signal of short duration and positive sign; (2) a similar signal of negative sign; (3) a signal of greater strength or duration, and of positive sign; (4) a signal similar to that last mentioned, and negative in sign. Of course an alphabet in which four signs take part can be composed of a much shorter series of combinations, than one in the construction of which only two signs take part.

The high state of insulation and conductivity of the cables is such, however, that it has not been found necessary to use any complicated form of sending apparatus such as the curb-key; but the ordinary recording keys are used instead, with a set of condensing plates in circuit.

At first the cables were worked slowly and at a rate of about eight words per minute, but this soon improved as the staff became more accustomed to the apparatus, and has steadily increased up to fifteen, and even seventeen words per minute on each cable.

In order to reduce the effect of earth currents (or those resulting from terrestrial magnetism), which frequently flow strongly in various directions and interfere somewhat with signalling, the earth connections of the cables have been carried several miles out to sea by means of short independent lengths of cable. The more equable temperature and small magnetic variation of the sea have thus diminished greatly the earth currents passing through the cables.

137. The charge for transmission was at first restrictive, the company being fearful of blocking up the cables. For three months the tariff was 20*l.* for a message of 20 words; but not withstanding this almost prohibitory rate of 1*l.* per word, several messages exceeding 800 words each were sent; one conveying a report for the "New York Herald," of the fight for the Championship between Mace and Goss; and another a verbatim copy of the King of Prussia's speech to the German Parliament after the conclusion of peace with Austria. The birth of a son was also announced in the London Times, "by Atlantic telegraph," the happy event having taken place at New York a day or two before. The tariff was reduced on the 1st of November to 10*l.* per message of 20 words.

138. Upon the return of the *Great Eastern* and other vessels to England, public dinners were given to Messrs. Glass, Canning, Clifford, Anderson, Prof. Thomson, Moriarty, and others connected with the expedition; and her Majesty recognised their great services by conferring knighthood and other honours upon some of those most prominently engaged in the undertaking. Of so much importance, indeed, was the Atlantic telegraph considered, that it formed a prominent topic of the Queen's Speech upon proroguing Parliament, as the following extract shows:—

> Her Majesty has great satisfaction in congratulating the country and the world at large on the successful accomplishment of the great design of connecting Europe and America by the means of an electric telegraph. It is hardly possible to anticipate the full extent of the benefits which may be conferred on the human race by this signal triumph of scientific enterprise, and her Majesty has pleasure in expressing her deep sense of what is due to the private company which, in spite of repeated failure and discouragement, has at length for the second time succeeded in establishing direct communication between the two continents. Her Majesty trusts that no impediment may occur to interrupt the success of this great undertaking, calculated, as it undoubtedly is, to cement yet closer the ties which bind her Majesty's North American colonies to their mother country, and to promote the unrestricted intercourse and friendly feeling which it is most desirable should subsist between her Majesty's dominions and the great Republic of the United States.

139. Since the opening of the cables the Astronomer Royal arranged with the Electric and Magnetic companies for the use of a through wire from Greenwich to Valentia; and in concert with Dr. Gould carried on a series of experiments between the Royal Observatory and the station at Foilhommerum, and thence through the Atlantic cables to Newfoundland and the American stations, in order to determine the exact longitude of places upon the American continent, so as absolutely to verify or correct previous calculations. These experiments were successfully carried out during November, 1866.

140. To show how exceedingly small an electric charge may be made to produce signals through the Atlantic cables; during the experiments carried on by Dr. Gould at Valentia, Mr. Collett, the superintendent at Newfoundland, actually sent a message with a battery composed of a *copper percussion cap and a small strip of zinc, which were excited by a drop of acidulated water the simple bulk of a tear!*

PLATE X

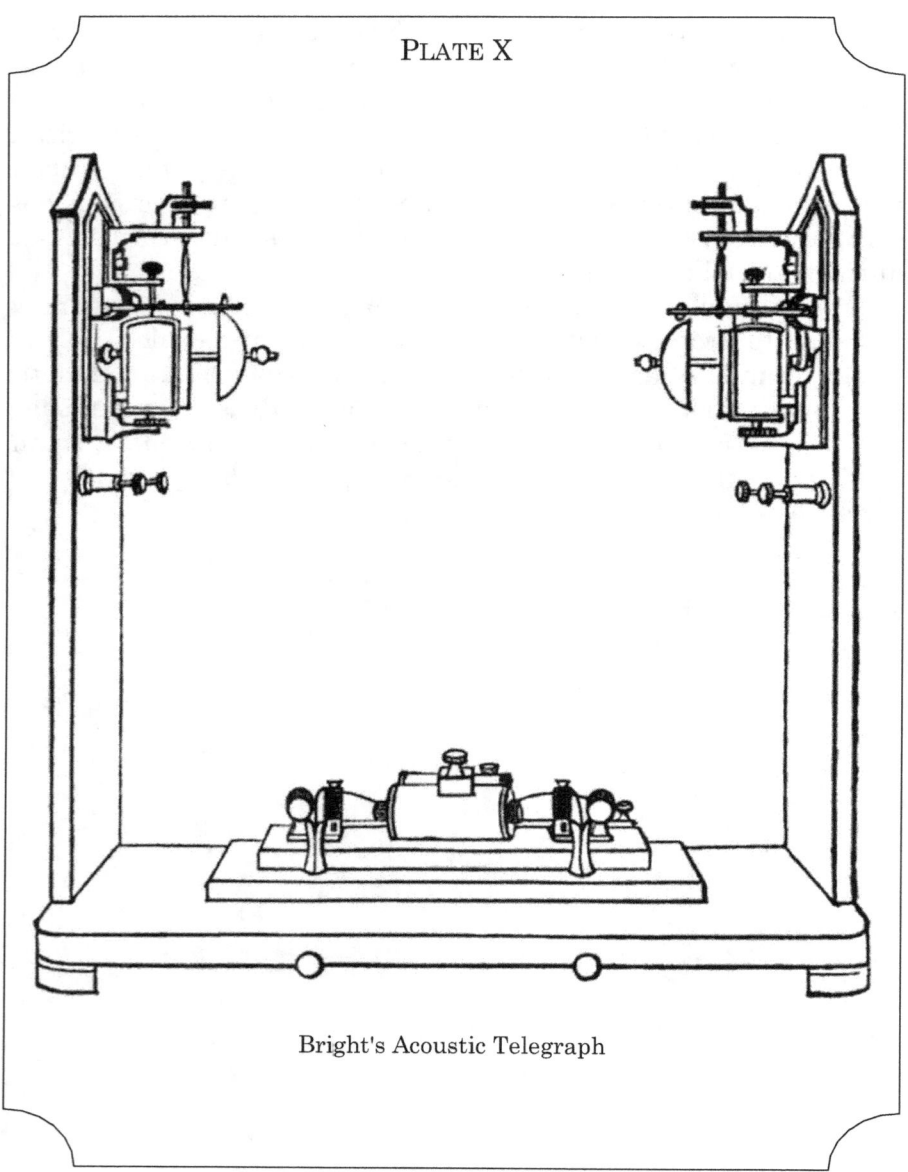

Bright's Acoustic Telegraph

10

SIGNALLING APPARATUS
NEEDLE AND ACOUSTIC INSTRUMENTS

141. IN entering upon a consideration of the various kinds of telegraphic instruments in use, it has appeared desirable to classify them; first, describing those producing signs by the combinations of movements of magnetised needles or beats of bells; secondly, apparatus making conventional marks upon paper; thirdly, instruments indicating the letters of the alphabet upon a dial, by the step-by-step movement of a pointer or index; fourthly, apparatus contrived to actually print the letters of the alphabet upon paper. The second, third, and fourth classes are necessarily much more complicated than the first, requiring escapements or trains of wheels to work them; and, therefore, the more simple forms are those first dealt with.

These different methods of communicating are applicable to the varied conditions under which they are required; those of a simple and inexpensive kind are specially adapted for railway purposes and minor stations, where the operators are railway porters or clerks not possessed of the amount of knowledge requisite for taking charge of and regulating the more complex apparatus in use on important circuits, where large numbers of messages have to be transmitted daily, and where there are skilled mechanics at hand to set right any derangement.

It may be taken as a rule that when simplicity is combined with a high speed of transmission and accuracy, the apparatus becomes the more capable of general application. The telegraphic instruments used to a large extent on the railways and at the minor stations in this country, are galvanometers, which make their signals by means of the deflections of magnetic needles, produced by the electric current.

These instruments are of two forms, the first, and most simple, consisting of one needle with its accessories, and the other of two independent needles, each accompanied by its own appendages. They are modifications of Messrs. Cooke and .Wheatstone's first telegraphs.

THE SINGLE NEEDLE INSTRUMENT.

142. This instrument consists of a galvanometer and a commutator, mounted in a case resembling in form and size that of an ordinary table time-piece.

A front view of it is given in Figure 72. On the upper part is a dial, in the centre of which the indicating needle appears, like the hand of a clock, fixed upon an axis. Its play to the right and left is limited by two ivory studs inserted in the face of the dial, a short distance on each side of its upper arm.

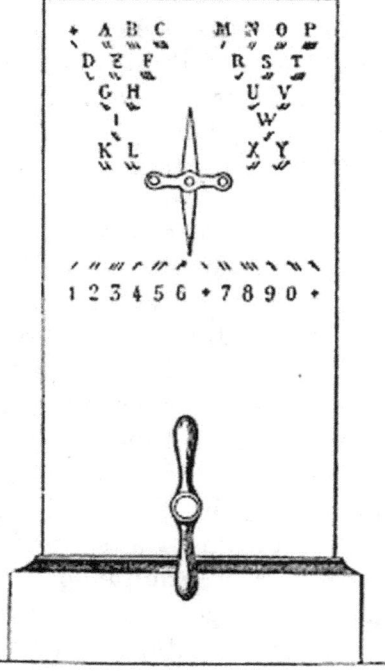

The handle which works the commutator, also fixed upon an axis, is presented at the lower part of the case, under the dial.

Upon the dial are engraved the letters of the alphabet, the ten numerals, and one or two arbitrary symbols, under each of which is engraved a mark, indicating the motions of the needle, by which the letter or figure is expressed.

The galvanometer is attached to the back of the dial, the axis of its magnetic needle passing through the dial, and carrying the indicating needle in front.

Fig. 72
Cooke and Wheatstone
single needle telegraph

The latter is also usually magnetic, its poles being reversed in their direction with relation to those of the interior needle, the effect of which is, that the current transmitted through the galvanometer has a tendency to deflect both needles in the same direction. The indicating needle, however, need not be magnetic. If it be sufficiently light, being free from magnetism, it will be carried by the axis to the right or left against the studs, by the deflections of the galvanometric needle which plays within the coils of the galvanometer, to which it is always parallel.

In connection with the instrument there are, as usual, an alarum and a galvanic battery.

By the commutator, the current produced by the battery may be transmitted upon the line-wire, or suspended or reversed in its direction, according to the position given to the handle. If the handle be vertical, as represented in the figure, the current is suspended, the arrangement of the commutator being then such as to cut off all communication between the battery and the line-wire. If the upper arm of the handle be turned to the right, the battery will be connected with the line-wire, on which accordingly the current will be transmitted. If the upper arm be turned to the left, the battery will still be connected with the line-wire, but with its poles reversed, so that the direction of the current on the line-wire will be reversed.

To comprehend the practical operation of the instrument, we are to consider that similar instruments, with similar accessories, are placed at each of the stations between which despatches are to be transmitted. To render the explanation more clear, let *S* and *S'*, Figure 73, be the two stations, *O* and *O'* the dials, *C* and *C'* the handles of the commutators, and *B* and *B'* the galvanic batteries. If it be intended to send a despatch from *S'* to *S*, the arm of the commutator, *O*, is left in its vertical position, so that no current can pass from the battery, *B*, to the line-wire, *L*. When the arm of *C'* is vertical, no current can pass from *B'* to *L*, and consequently the needle of *O* will remain in the vertical direction, without deflection. If the upper arm of *C'* be turned to the right, *r,* the current from *B'* passing along *L*, will flow through the coil of the galvanometer at *S*, and will deflect the indicating needle to the

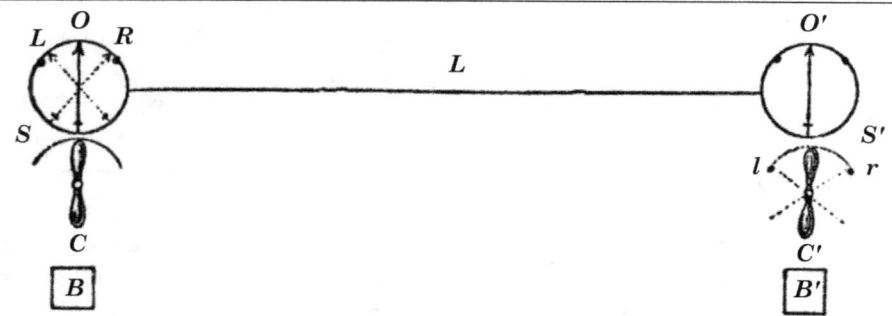

Fig. 73
Simultaneous movements of connected needle instruments

right, so that it will lean upon the right hand stud, **R**. If **C'** be then turned back to the vertical direction, the current will be suspended, and the needle at **S** will return to the point **O**. If the upper arm of **C'** be then turned to the left, **l**, the current will be again transmitted upon the line-wire, **L**, but in a direction contrary to its former course; and thus passing through the galvanometer at **S**, in a contrary direction, the needle, which was before deflected to the right hand stud, **R**, will now be deflected to the left hand stud, **L**.

Thus, it appears that, according as the upper arm of **C'** is turned to the right or left, or placed in the vertical position, the needle on the dial at **S** is also turned to the right or left, or placed in the vertical position. In a word, whatever position is given to the handle of the commutator at **S'**, a corresponding position is assumed by the indicating needle at **S**, and these changes of position of the indicating needle at **S** are absolutely simultaneous with the changes of position of the handle of the commutator at **S'**.

The manner of expressing the letters and figures is by making repeated deflections of the needle right and left, making a short pause at the end of each letter-signal. Thus, two deflections to the left express A; three, B; four, C; while one expresses the completion of a word. One to the right expresses M; two, N; three, O; and four, P. In the same manner, L is expressed by four deflections, which are, successively, right, left, right, and left.

As these signs are purely arbitrary, and may be changed in every independent telegraph, it is not necessary here to notice them further.

Besides the signals which express letters and figures, it is usual to adopt others to express words or phrases of very frequent occurrence, such as, "I don't understand," "I understand," " Wait," " Go on," " Repeat," &c.

It is usual, though not necessary, for the agent who sends a despatch, to pass the current through his own instrument, so that his indicating needle shows exactly the same deflections as the indicating needle of the station he addresses. Thus, when *S'* addresses *S*, his own indicating needle, *O'*, speaks as well as the indicator, *O*, of the station, *S*.

All that has been stated of the transmission of the same despatch through a series of stations, of cutting off the transmission from all stations except that to which it is exclusively addressed, of the use of the alarum, &c., is applicable, without any important modification, to this form of telegraphic instrument.

THE DOUBLE NEEDLE TELEGRAPH.

This is nothing more than two single needle telegraphs, such as has been just explained, mounted in the same case, their indicating needles playing side by side upon the same dial, and the handles of their commutators placed so that they can be conveniently worked at the same time, by the right and left hand of the telegraphic agent. Each instrument is altogether independent of the other, having separate accessories, and transmitting its current upon a separate line-wire; and, therefore, twice the expenditure in the construction and maintenance of wires is required in the use of the double-needle apparatus, as compared with those instruments which are worked through a single wire; while it has been found in practice that the increase of speed obtained from the use of two wires in one instrument is not proportionately greater. For this reason, the use of the double-needle apparatus has much decreased, and would probably be altogether discontinued but for many of the railway employees being accustomed to the interpretation of its signals. It is a far more difficult instrument to learn than the single-wire telegraphs, owing to the necessity for the movements of *both* needles to be noted at the same time by the eyes of the clerk.

The purpose of this form of instrument is merely to accelerate the transmission of despatches, by enabling the agent to produce the signals expressing letters and figures in more rapid succession. In the single instrument there are only two signs made by one deflection of the needles, viz., a deflection to the right and one to the left. In the double instrument there are eight such signs, viz., two with each needle, as in the single instrument, and four obtained by combining the deflections of the two needles. Thus, if *O* express the position of the needle without deflection, *r* a right hand, and *l* a left hand deflection, and *R* the right hand, and *L* the left hand needle, the following eight signals may be made in the time of a single motion of either needle, as represented in the table below:–

L	R
r	O
l	O
O	*r*
O	*l*
r	*r*
l	*l*
r	*l*
l	*r*

With a single needle two deflections can only make four signals, viz., *r r, l l, r l, l r*. But with two needles, these being combined with single deflections and with each other, a greater number of different signals can be obtained than are sufficient to express the letters and numerals, each being made in the time necessary for two deflections of a single needle. A front view of a double-needle telegraph is given in Plate I, p. 6.

The small case at the top contains the alarum, and the small handle at the side of the large case is the commutator by which the current is turned on and off the alarum. The two large handles which appear in front are those of the commutators, which produce the changes of direction of the current, and when inclined to the right or left the needles acted on by the current assume a like position.

143. A modification of Messrs. Cooke and Wheatstone's single needle instrument was introduced by the late Mr. Highton, which was brought into considerable use. In his apparatus a circular magnet and coil were employed, in order to reduce the vibration arising after the movement of the long needles adopted by Messrs. Cooke and Wheatstone.

Fig. 74
Henley and Foster magneto-electric telegraph

144. The British and Irish Magnetic Telegraph Company, at first retaining the needle-indicators generally used in England, employed the magneto-electric in place of the voltaic current on many of their lines. The instruments they adopted are those which were patented by Messrs. Henley and Forster, with some modifications introduced by Messrs. Bright. This form of telegraph is shown in Figure 74.

The current is produced by electro-magnets, whose poles are moved in close proximity with those of strong compound permanent magnets. These latter are represented at *A* (Figure 74). At their poles a straight piece of soft iron is placed, by the inductive influence of which the magnetism of the several bars composing the compound magnet is collected and combined. The electro-magnets are formed in the usual way, and are mounted on centres on which they are turned by levers, which project from either side of the case, so that the agent can work one with each hand.

When these levers are pressed down, the electro-magnets are reversed in the relation of their poles to those of the permanent magnets, and momentary currents are transmitted on the conducting wires; and when the levers are observed to rise to their former position, momentary currents are again transmitted, but in a contrary direction.

THE ACOUSTIC TELEGRAPH.

145. The ordinary double-needle telegraph, described at the commencement of this chapter, has given way to the Morse recording instrument on all the important lines of the Electric Company; and, in like manner, the acoustic telegraph, an improved method of signalling by sound, instead of by visual signals, introduced in 1854 by Messrs. Bright, has superseded the double-needle magnetic instrument on the British and Magnetic wires. Dr. Noad, in his "Manual of Electricity," refers to this improvement as follows:—

Under the ordinary system of telegraphing, it is necessary to employ a transcriber to write down the words as interpreted from the visual signals and dictated to him by the receiving operator, whose eyes being fixed on the rapidly-moving needles, could not be engaged in conjunction with his hands in writing. It was found that, owing to the frequent occurrence of words of nearly similar sound, the transcriber sometimes unavoidably misunderstood the meaning of the receiving operator, and altered the sense of the despatch by writing the wrong word. Such words as *two*, *too*, *to*; *four*, *for*; *hour*, *our*; may, for instance, be very easily confounded.

Such errors cannot, however, arise when the clerk who has. heard each word pass through the acoustic telegraph, letter by letter, is able, from his eyes being at liberty, to write what he has received without the aid of an amanuensis. Besides the saving in staff and in mistakes, any injury to the eyes of the clerks is prevented, and an appeal is made to an organ far better capable of endurance and accurate interpretation.

The acoustic apparatus used is very simple, consisting of a hammer in connection with a lever, which is acted upon by every polarisation of a set of electro-magnets by the local current, and thereupon strikes a small bell. A pair of these bells are connected to each wire; one bell is struck upon the passage of the positive, and the other of the negative current, the alphabet being readily formed by the difference in their tone and the number of beats.

The nature of this apparatus is shown in Figure 75, where *A* represents the hammer of the bell, held back to a stop by a flexible spring. The rod of the hammer is fixed to the projecting horns of the moveable soft iron core of an electro magnet *B'*. The electro-magnet *B'* is fixed opposite to a horseshoe electro-magnet *B*; and the connections are so arranged that on the current passing from the relay, the electro-magnets are polarised with their opposite poles to one another. Upon a current passing, the bell is at once struck, and the bell being muffled so as to produce a short sound, the blow may be repeated as rapidly as desired without any vibration from one sound interfering with that succeeding it. One bell is placed on one side, and one on the other of the clerk.

A local battery supplies the mechanical power required to strike the bells. The battery is put in connection with either bell

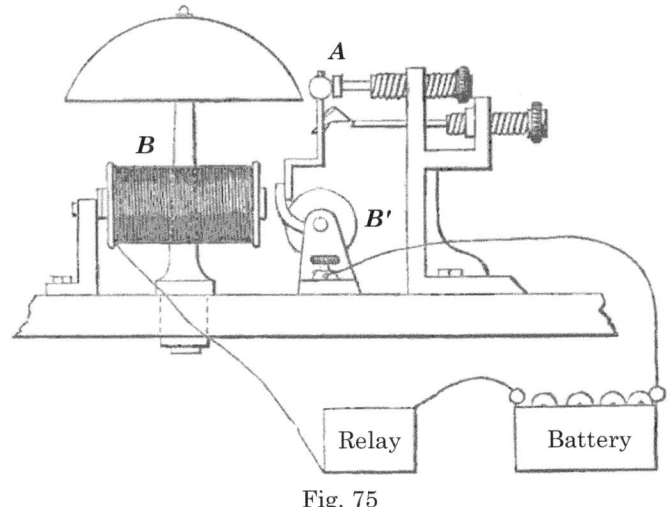

Fig. 75
Acoustic telegraph instrument

according to the current, positive or negative, passed through a relay, the arrangement of which is shown in Figure 75 The keys with which currents are sent to work this apparatus are of a simple commutating form. By pressing down one key the current is made to pass in one direction, and in the reverse when the other key is used.

This form of telegraph, like the Morse instruments, only requires one wire; but in point of speed it has a great advantage, as it utilises both positive and negative currents, while the Morse is only available for one current. The alphabet of the Morse system is composed of dots and long marks, the long mark being produced by holding on the current with the sending key about three times the length of the dot; but in the acoustic telegraph the two signs, requisite in their combination to form the alphabet, are produced by beats on the two differently pitched bells; each beat being the equivalent in time to a dot on the Morse instrument, so that the acoustic alphabet only occupies about half the time of the Morse.

This is found to be practically arrived at on the busy lines of the Magnetic Company, where a speed of 35 to 40 words per minute is frequently attained on long despatches transmitted upon a single wire by the acoustic apparatus. In forming the alphabet the most simple combinations of beats are used to express those letters most frequently required.

146. When the establishment of lines of electric telegraphs was proposed in France, the old aerial telegraph was, and had been for more than half a century, in operation, and formed a department in the public administration of considerable importance, employing an extensive body of agents, dispersed throughout the country, most of whom were specially instructed and qualified for the business.

The commission appointed by the government required that the electro-telegraphic instruments should exhibit the same signals as had been already used in the case of the former telegraph.

The old telegraph consisted of a long straight bar, R, R', Figure 76, called a regulator, to the extremities of which two shorter bars, r r', called indicators, were attached by pins or pivots, so that each indicator was capable of turning on its pivot, so as to make any desired angle with the regulator.

Fig. 76
French aerial telegraph indications

If we suppose the circle described by each indicator to be divided into eight equal arcs of 45°, and that any convenient mechanism is provided, by which the agent who conveys the signals can at will give to each indicator any of these eight positions, each indicator would be capable of making eight signals, and by combining these in pairs, the two indicators worked together would be capable of giving sixty-four signals. It is evident that even this large number of signals might be further multiplied, by giving to the regulator itself a motion round its centre, so that it might at will assume the horizontal or vertical position, or might take an intermediate direction.

In transferring this system of signals to the electric telegraph, the regulator is supposed to be placed permanently horizontal, and the two indicators to be capable of receiving any of the eight positions here explained.

The telegraph contrived by M. Breguet, to exhibit such a system of signals, consists, like the double needle telegraph, of two distinct and perfectly similar instruments, one for each of the indicators. They are mounted side by side with their accessories in the same case, at a distance apart sufficient to allow the indicators to revolve without mutual obstruction, and sufficiently near each other to allow the same person to work both at the same time with his right and left hand. Each instrument consists of an indicating apparatus and a commutator.

A view of the indicating apparatus is given in Figure 77. The two indicators are fixed upon axes placed in the same horizontal line upon the dial. These axes, passing through the dial, carry behind it two escapement wheels, which are controlled by two anchors acted upon by the armatures of two electro-magnets, from which they receive vibrations, like those of a pendulum. The escapement wheels are impelled by the force of two main-springs,

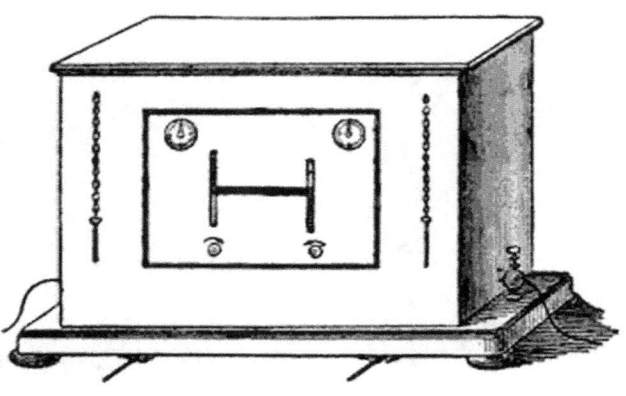

Fig. 77
Breguet receiving apparatus

transmitted to them by two similar trains of clock-work.

Thus, for each swing of the anchor, the indicator makes one motion forward, and as the escapement wheels have each only four teeth at equal distances, one complete revolution of these wheels must cause the indicators to make a complete revolution by eight distinct motions, produced by the four swings of the anchor to the right, and the four swings to the left.

During a revolution of each of the escapement wheels, therefore, each of the indicators takes successively the eight positions required in the proposed system of signals, and since the motions of the indicators are governed by the anchors, those of the anchors by the armatures of the electro-magnets, and those of the electro-magnets by the successive pulsations of the electric current, it follows that if it can be contrived that commutators at one of the stations shall govern the pulsations of the current at the other, they will necessarily govern the motion of the indicators at that other station.

At the upper corners, right and left of the front of the case, are two dials, in the centre of which are axes, which act, when turned, upon the springs which draw back the armatures of the two electro-magnets, and near them keys for their adjustment are suspended by chains. The springs are raised or relaxed, according as the keys are turned in the one direction or the other.

Under the indicating arms are two axes with square ends, by which the two systems of clock-work can be wound up, which is done by the same keys.

It remains, therefore, to show the manner in which the pulsations of the current are governed by the commutator. One of the commutators is represented in Figure 78.

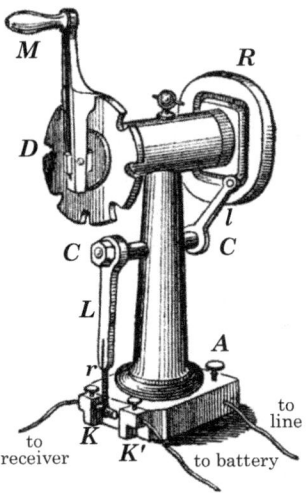

Fig. 78
Breguet sending apparatus

The handle *M* is fixed upon an axis which turns in the centre of a fixed disc *D*, the edge of which is divided into eight equal parts by small notches. A short pin projects from the handle which falls successively into these notches, but which can be withdrawn from them when it is required to turn it. On the remote end of this axis a disc *R* is fixed, which turns with it, in the face of which a square groove is cut, rounded at the corners, in which a pin projecting from a short lever *l* is moved. This lever *l* is fixed on the axis *C C*, upon the other end of which is fixed the lever *L*, the lower end of which carries a small piece of metal *r*, which, when the lever vibrates right and left, is thrown alternately against the contact-piece *K* and *K'*.

Supposing that the commutator is placed at the station *S*, the line-wire which comes from the station *S'* enters the foot, and is held there by a tightening screw *A*. This wire is in metallic connection, through the pillar, with the lever *L*, and consequently with the piece of metal at its lower end, which oscillates between the contact-pieces *K* and *K'*. This piece of metal, *r*, may therefore be considered as virtually the extremity of the conducting wire between the stations *S* and *S'*.

Attached in like manner, by tightening-screws, to the two contact-pieces *K* and *K'* are two wires, one of which is connected with the battery, and the other with one end of the coil-wire of the electro-magnet, in the indicating instrument of the station. The other end of this coil-wire is either connected with the line-wire which proceeds to the succeeding station, or with the earth, at the option of the agent, a commutator being provided by which this change of direction may be made.

The arrangement of the apparatus is such, that when the handle **M** of the commutator is presented vertically upwards, as represented in the figure, the pin being in the highest notch, the lever **L** presses against the contact-piece **K**.

Let the highest notch be supposed to be numbered 1, and the others proceeding round the disc, in the direction of the motion of the hand of a clock, be numbered successively 2, 3, 4, 5, 6, 7, and 8.

The arm **M** being, as represented in the figure, in the notch 1, let it be moved to the notch 2. The lever **L** being moved to the right, the piece **r** will be thrown upon **K'**. Being then in connection with the battery-wire, the current will pass by **r** and **L** to **A**, and thence by the line-wire to the corresponding point of the commutator at the distant station, and thence through the pillar to the lever **L'** and the piece **r'**. But since, as has been just explained, **M'** is in the notch 1', the piece **r'** must rest against **K**. The current, therefore, arriving at this point, will pass from **K** by the wire to the coil of the distant instrument to which it will impart magnetism, so that it will attract the armature, and move the anchor of the escapement, so as to make the indicator move from the vertical position 45° in the direction of the hand of a clock.

If the handle **M** be now moved from notch 2 to notch 3, the lever **L** will be thrown back to **K**, and the contact with **K'** being broken, the current will be suspended, and the electro-magnet at **S'** losing its power, the armature will recoil from it by the action of the spring, and the anchor of the escapement being again moved, the indicator will be advanced through another angle of 45°, and will be then in the horizontal position pointing to the right.

In like manner, it may be shown that when the arm **M** is moved from the notch 3 to the notch 4, the indicator will be moved from the horizontal position to one which will make an angle of 135°, with its original direction, or what is the same, 45°, with the position in which it would point directly downwards.

From what has been explained, the process necessary, as well for receiving as for transmitting a despatch will be understood. In the reception of a despatch, the agent has only to place the handle of his commutator in notch 1, and to see that his indicator is vertical. After that he has only to observe the successive attitudes assumed by the two indicators upon the dial before him, and to write down the letters they successively express.

Since this form of telegraph gives 64 signs, while 26 are sufficient for the alphabet, and 10 for the numerals, there are 28 signs disposable for abridgments, such as syllables, words, and phrases of most frequent occurrence.

The battery employed in working these telegraphs is at present invariably that of Daniel. Formerly Bunsen's battery was used at chief stations, where great power is often required, but this has now been discontinued.

Between the point **K'** and the battery a commutator is placed, by means of which the agent can bring into action a greater or less number of the pairs composing the battery, so as to proportion the power to the distance to which the current is to be transmitted, or to the resistance it may have to overcome.

PLATE XI

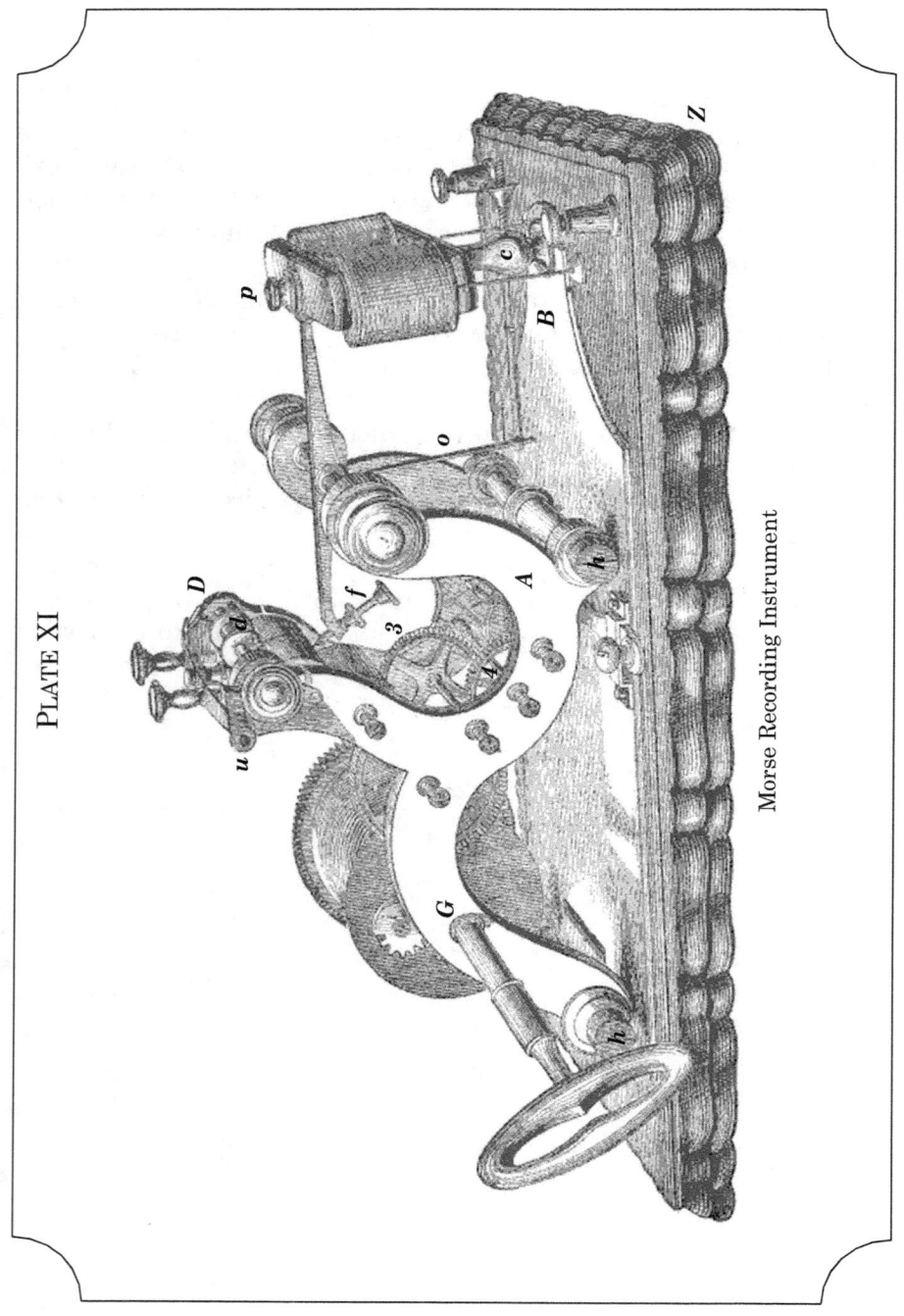

Morse Recording Instrument

11

SIGNALLING APPARATUS
RECORDING INSTRUMENTS

MORSE'S TELEGRAPH.

147. WE now come to the recording telegraphs, one of which, that invented by Prof. Morse, and the first telegraph of the United States, is far more extensively used than any other instrument; having been adopted generally in Europe, and to a considerable extent in England, besides being nearly exclusively used throughout the United States.

Although it utilises the electric current in one direction only instead of in both, and is not therefore so quick in signalling as apparatus in which both currents are turned to account, yet the simplicity of arrangement and strength of mechanism of this instrument are such that it has become a general favourite.

A diagram of the instrument in its most usual form is given in Figure 79.

M is the electro-magnet; *H* is an armature working on the centre *c*; *i* an adjusting screw to limit the play of the armature, and prevent its contact with the electro-magnet at *p*; *d* another adjusting screw to limit its play in the other direction; *t* a metallic style which marks by pressure a band or ribbon of paper drawn from the roll *R*, and carried between the rollers *o* and *o'*; *P* the ribbon of paper discharged from the rollers *o o'* after being impressed by *t* with the telegraphic characters; *I*, *B*, &c, clockwork from which the rollers *o o'* receive their motion, by which motion the ribbon of paper is drawn from the roller *R* ; *f* the spring which draws the arm *H* of the electro-magnet from the armature; *S S*

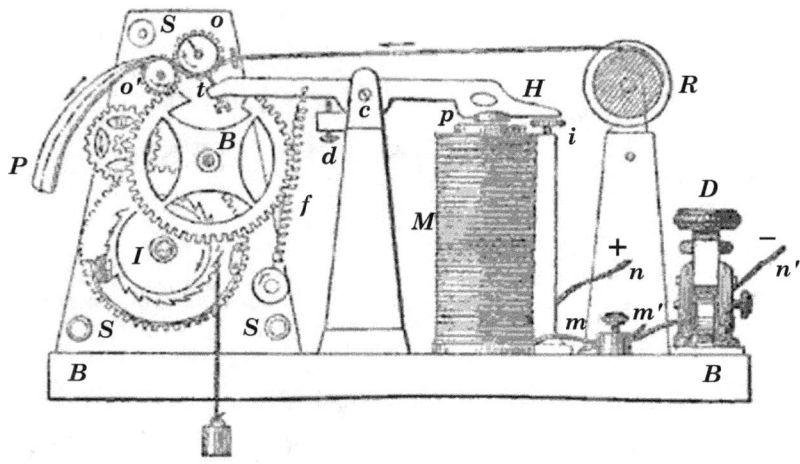

Fig. 79
Morse's American recording telegraph

the upright pieces supporting the clockwork; **B B** the base supporting the instrument; **D**, the key commutator, by which the current transmitted along the line-wire is alternately transmitted and suspended; **m, n, m' n'**, wires by which the coil of the electro-magnet and the poles of the station battery are put in connection with the line-wires.

The general principle of this and all similar apparatus has been already so fully explained, that little more need be said here to render it intelligible. If it be desired to transmit a despatch to a distant station, the battery at the transmitting station is put in communication with the line-wire, and by the action of the key **D** the current is alternately transmitted and suspended during longer and shorter intervals, which are determined by the conventional telegraphic letters. The action of the style **t** against the ribbon of paper which passes over it at the station receiving the despatch, corresponds exactly with the action of the key **D** at the station from which the despatch is transmitted; and combinations of longer and shorter marks or lines and dots are produced upon the ribbon of paper by its pressure, as is shown in the figure.

The particular combinations of lines and dots used to express the letters are obviously arbitrary. As a matter of convenience and means of expedition, the letters of most frequent occurrence are expressed by the most simple signs, and consequently the selection of signs for the different letters will vary with the language in which the despatch is expressed.

The following are the telegraphic characters usually adopted in the Morse code:–

				Numerals.	
A ·—	J —·—·	S ···	1 ·———·	9 —··—
B —···	K —·—	T —	2 ··—··	0 ————
C ·· ·	L ——	U ··—	3 ···—·	
D —··	M ——	V ···—	4 ····—	
E ·	N —·	W ·——	5 ————	
F ·—·	O ··	X ·—··	6 ······	
G ——·	P ·····	Y ··—··	7 ——··	
H ····	Q ··—·	Z ··· ·	8 —····	
I ··	R · ··	& · ···		

This telegraphic apparatus being that which has been by far the most extensively brought into use, it may be useful here to present the instrument and its appendages in the form in which it has been most recently constructed in the United States, which has been recommended by the American telegraphic confederation, as being that which it would be most advantageous to adopt generally, so that all the parts being manufactured of the same pattern and size no difficulty would be found in replacing any of them in case of fracture.

A perspective view of the instrument, omitting the paper roller and ribbon, is given in Plate XI, p. 184.

- **Z**. The wooden base upon which the instrument is screwed.
- **B**. The brass base plate attached to the wooden base **Z**.
- **A**. The side frames supporting the mechanism.
- **h**, **h**. Screws which secure the transverse bars connecting the side frames.
- **G**. The key for winding up the drum containing the main-spring, or supporting the weight according as the mechanism is impelled by one or the other power.

- *3, 4.* Clock-work.
- *u.* A lock or gauge to regulate the pressure of the rollers on the paper.
- *c.* The pillar supporting the electro-magnet.
- *p.* The adjusting screw passing into the pillar, *c*, projecting through the armature, to enable the telegraphist to adjust the sound of the back stroke of the armature at pleasure.
- *o.* The spring bar, and
- *d*, the screw to adjust the action of the pen lever.
- *D.* The apparatus for adjusting the paper rollers.
- *f.* The adjusting screw of the pen lever.

The form of the relay magnet recommended, is given in Figure 80.

- *A B.* are the helices or coils.
- *C.* The supporter of the magnet lightly screwed to
- *W*, the connecting bar of the magnets.
- *Y.* Rosewood or ivory ends of magnets.
- *D.* Armature screwed to
- *E*, an upright lever;
- *F*, its axis, surrounded by a spiral spring, to perfect the connection in case of a fault at the ends of the axle.
- *M.* The spring to produce the recoil of *D* and *E*.
- *L.* Its adjusting screw.
- *H.* An adjusting screw to limit the play of *E* towards the magnet,

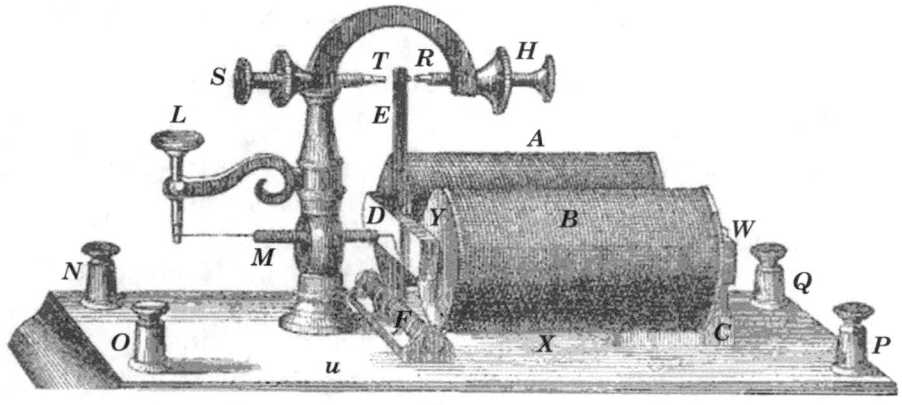

Fig. 80
Relay for Morse instrument

- **R**, its point of platinum.
- **S**. An adjusting screw to limit the play of **E** from the magnet.
- **T**. Its insulating point, in ivory.
- **O N**. Screws to connect with the wires of the station battery.
- **P Q**. Screws to connect with the line wires.
- **X**. The point where the coil wire passes through
- **u**, the base of the magnet.

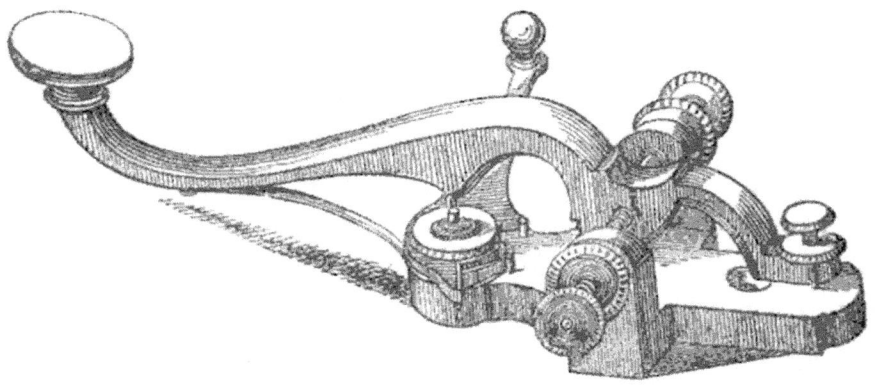

Fig. 81
Morse commutator key

The form recommended for the key commutator is represented in **Figure 81**. When the key is held down the circuit is perfect. It is not liable to wear and to produce a doubtful connection. The whole arrangement is designed to avoid the evils heretofore existing, and perfect every questionable part. The anvil of the key is well made, firm, and capable of hard wear, regardless of the adjustment of the key lever. The hammer of the key lever is also firm, and made of good platina wire, and securely made fast in the key lever. The adjusting screws of the axle are arranged according to the best mode, to secure the most perfect action. The elevation of the key lever can be adjusted to suit the operator, by elevating the key frame, or otherwise.

Inking apparatus has been introduced, by which the marks upon the paper ribbon, instead of being mere indents, are blackened by a disc of carbon or inking roller.

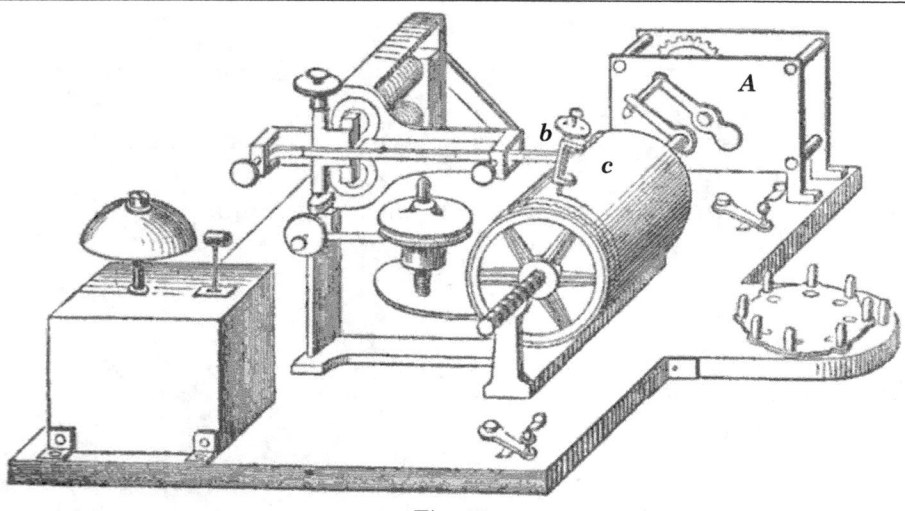

Fig. 82
Froment's recording instrument

FROMENT'S WRITING TELEGRAPH.

148. This apparatus is represented in Figure 82.

The paper upon which the telegraphic characters are written is rolled upon the surface of a drum *c*. The pencil *b* is pressed by a spring upon the paper. The drum is made to revolve by clockwork in the usual manner contained in the case *A*. If the paper be moved without moving the pencil, the latter will trace a straight line; but if the pencil be moved to and fro by the action of the electro-magnet and recoil spring, a zigzag line will be formed by the vibrations imparted to the pencil by the magnet, or what is the same, by the pulsations of the current.

To equalise the wear of the pencil, a slow motion of rotation is imparted to it by wheels adapted for that purpose.

The commutator by which the pulsations which determine the signals are produced, is a wheel, at the circumference of which are five metallic divisions with intermediate spaces vacant, so that in each revolution the current is transmitted five times, and suspended five times. If it be desired to produce a single pulsation, the wheel is moved through the fifth part of a revolution; if it be desired to produce three pulsations it is moved through three-fifths of a revolution, and so on. For each pulsation, one zig-zag is made by the pencil at the station to which the despatch is transmitted.

The signs adopted in this telegraph to express the letters, are various numbers and combinations of zigzag forms.

This apparatus has been superseded.

BAIN'S ELECTRO-CHEMICAL TELEGRAPH.

149. The manner in which the decomposing power of the current is capable of producing written characters at a distance from the hand of the writer has been already explained.

Of the forms of telegraph in which this principle is brought into play, the only one which has been practically applied on an extensive scale is that projected by Mr. Alexander Bain, and shown in Figure 83.

To render this instrument understood, let us suppose a long strip of paper to be wetted with a solution of prussiate of potash, to which a little nitric and hydrochloric acid have been added. Let a metallic roller be provided corresponding in magnitude with the

Fig. 83
Bain's electro-chemical telegraph

sheet of paper, and in communication with a galvanic battery so as to form its negative pole. Let a piece of steel or copper wire forming a pen be put in connection with the same battery so as to form its positive pole. Let the strip of moistened paper be now laid upon the metallic roller, and let the steel or copper point which forms the positive pole of the battery be brought into contact with it. The galvanic circuit being thus completed, the current will be established, the solution with which the paper is wetted will be decomposed at the point of contact, and a blue or brown spot will appear. If the roller be now moved with the paper under the pen, the continuous succession of spots will form a blue or brown line, and characters may be thus written upon it as it were in blue or brown ink.

An extremely feeble current is sufficient to produce this effect; but it will be necessary, when the strength of the current is very much reduced, to move the roller more slowly, so as to give the time necessary for the weakened current to produce the decomposition. In short, a relation exists between the greatest speed of the roller which is capable of leaving a mark, and the strength of the current; the stronger the current, the more rapidly may it be moved.

Let us suppose that the galvanic circuit is completed in the manner customary with the electric telegraph, that is to say, the wire which terminates at the point of the electro-chemical pen is carried from the station of arrival to the station of departure, where it is connected with the galvanic battery, and the returning current is formed in the usual way by the earth itself. When the communication between the wire and the galvanic battery at the station of departure is established, the current will pass through the wire, will be transmitted from the point of the electro-chemical pen to the moistened paper, and will make a blue or brown line on this paper. If the current were continuous and uninterrupted, this line would be an unbroken spiral; but if the current be interrupted at intervals, during each such interval the pen will cease to decompose the solution, and no mark will be made on the paper. If such interruption be frequent, the spiral, instead of being a continuous line, will be a broken one, consisting of lines interrupted by blank spaces. If the current be allowed to act only for an instant of time, there will be a blue or brown dot upon the paper; but if it be allowed to continue during a longer interval, there will be a line.

Now, if the intervals of the transmission and suspension of the current be regulated by any agency in operation at the station of departure, lines and dots corresponding precisely to these intervals will be produced by the electro-chemical pen on the paper, and will be continued regularly along the spiral line already described. It will be evident that characters may thus be produced on the prepared paper corresponding to those of the telegraphic alphabet already described in the case of Morse's telegraph, and thus the language of the communication will be written in these conventional symbols.

There is no other limit to the celerity with which a message may be thus written, save the sufficiency of the current to effect the decomposition while the pen passes over the paper, and the power of the agency used at the station of departure to produce, in rapid succession, the proper intervals in the transmission and suspension of the current.

The succession of intervals of transmission and suspension of the current on which the production of the written characters on the prepared paper depends, may obviously be produced by the key commutator; and with that instrument at the station from which the despatch is transmitted, an agent can convey in the same manner and with the same celerity as in the case of the telegraph of Morse, or that of Froment; and such is in fact the manner in which despatches are transmitted with this apparatus.

We shall now explain the means by which a greater celerity may be obtained. The despatch must pass through the following preparatory process:–

A narrow ribbon of paper is wound on a roller, and placed on an axis on which it is capable of turning so as to be regularly unrolled. This ribbon of paper is passed between rollers under a small punch, which striking upon it makes a small hole at its centre. This punch is worked by a simple mechanism so rapidly, that when it is allowed to operate without interruption on the paper passing before it, the holes it produces are so close together as to leave no imperforated space between them, and thus is produced a continuous perforated line. Means, however, are provided by which the agent who superintends the process, can, by a touch of the finger, suspend the action of the punch on the paper, so as to allow a longer interval to elapse between its successive strokes upon the paper. In this manner a succession of holes are perforated in the ribbon of paper, separated by imperforated spaces. The manipulator, by allowing the action of the punch to continue uninterrupted for two or more successive strokes, can make a linear perforation of greater or less length on the ribbon, and by suspending the action of the punch these linear perforations may be separated by imperforated spaces.

Thus it is evident, that being provided with a preparatory apparatus of this kind, an expert agent will be able to produce on the ribbon of paper as it unrolls, a series of perforated dots and lines, and that these dots and lines may be made to correspond with those of the telegraphic alphabet already described.

Let us now imagine the message thus completely inscribed on the perforated ribbon of paper. This ribbon is again rolled as at first upon a roller, and is now placed on an axle attached to the machinery of the telegraph. The extremity of the perforated ribbon **B** at which the message commences is now carried over a metallic

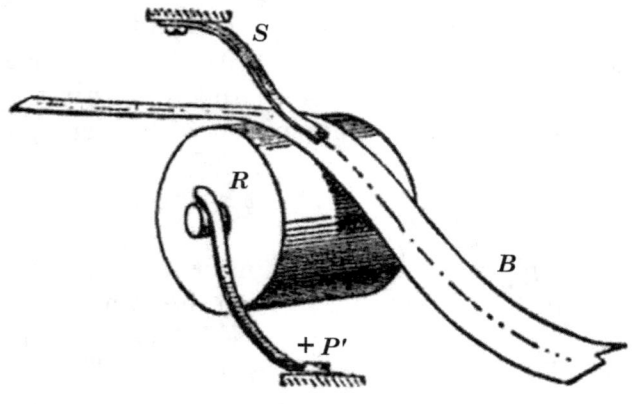

Fig. 84
Bain's electro-chemical telegraph transmitting ribbon

roller **R**, which is in connexion with the positive pole **P'** of the galvanic battery. It is pressed upon this roller, as represented in Figure 84, by a small metallic spring **S**, terminating in points like the teeth of a comb, the breadth of which is less than that of the perforations in the paper. This metallic spring is connected with the conducting wire which passes from the station of departure to the station of arrival. When the metallic spring falls into the perforations of the ribbon of paper as the latter passes over the roller, the galvanic circuit is completed by the metallic contact of the spring with the roller; but when those parts of the ribbon which are not perforated pass between the spring and the roller, the galvanic circuit is broken and the current is interrupted.

A motion of rotation, the speed of which can be regulated at discretion, is imparted to the metallic roller by clockwork or other means, so that the ribbon of paper is made to pass rapidly between it and the metallic spring, and, as it passes, this metallic spring falls successively into the perforations on the paper. By this means the galvanic circuit is alternately completed and broken, and the current passes during intervals corresponding precisely to the perforations in the paper. In this manner the successive intervals of the transmission of the current are made to correspond precisely with the perforated characters expressive of the message, and the same succession of intervals of transmission and suspension will affect the writing apparatus at the stations of arrival in the manner already described.

To the celerity with which the despatch can be written at the station of arrival there is no other limit than the time which is necessary for the electric current to produce the decomposition of the chemical solution with which the prepared paper is saturated.

It may be asked then why this form of telegraph, affording as it does the means of obtaining a celerity of transmission so far exceeding any other that has been projected, has not been universally adopted?

To this it may be answered that the celerity here described can only be attained after the despatch to be transmitted has been marked in the pierced telegraphic characters on the ribbon of paper, and that the process of so marking it would not be more rapid, however expert the operator might be, than that by which the same operator would transmit the same despatch directly by the key commutator, either with this telegraph or those previously described. If, therefore, the time necessary to commit the despatch in telegraphic characters to the perforated ribbon of paper, be included in the estimate of the time of its transmission from station to station, this form of telegraph is not only slower and consequently less efficient than the single needle, acoustic, or Morse, but is slower than any other form of telegraph whatever.

A practical difficulty has also been found in working the Bain instrument at a high speed, owing to certain intervals being required between the communication of successive currents to the conducting wires, and to the great amount of battery power required to make the marks on the chemical paper acted upon at the further end. This class of telegraph is also liable to more interference than others, from the circulation of currents of terrestrial magnetism along the wires, producing marks upon the paper independent of those intended to be communicated.

150. A modification of the electro-chemical telegraph has been contrived by Mr. Bakewell, by which a despatch may be transmitted to any distant station, and then delivered in the handwriting of the person who transmits it.

By this method, a person at any station, as for example at London, may write a communication in characters used in common writing or printing on paper placed at another distant station, as for example at Trieste, and this writing shall be traced on the paper with as much precision as if the person writing held the pen in his hand.

151. This apparatus has been improved by M. Caselli.

We may imagine that the electro-chemical pen placed on the paper at Trieste is extended to London, and there held and directed by the hand of the writer, for this it is which almost literally takes place. The conducting wire, in connection with that part of the electro-chemical pen which is held in the hand, which extends from Trieste to London, may be considered as only forming part of this pen, and the end of such pen at London, held and directed by the hand of the writer, will communicate a motion to its point at

Trieste, in exact correspondence with the characters formed by the hand of the writer.

Thus, if the writer at London move the extremity of the conducting wire so as to write a phrase and his usual autograph, the point at Trieste will there inscribe on the prepared paper the same phrase with the same signature annexed, and the writing of the phrase and the signature will be identical with that of the writer.

In the same manner a profile or portrait, or any other outline drawing, may be produced at a distance. The methods of accomplishing this depend, like the other performances of electricity in this application of it, on the alternate transmission and suspension of the current, and on its decomposing power; but as they are at present more matters of curiosity than of practical utility, we shall not detain the reader with any more detailed notice of them. This form of telegraph has proved too slow in operation, and too uncertain and difficult in management, to allow of practical application; besides which, it requires the apparatus to move onward at exactly the same speed at each end, a result almost impossible to attain.

PLATE XII

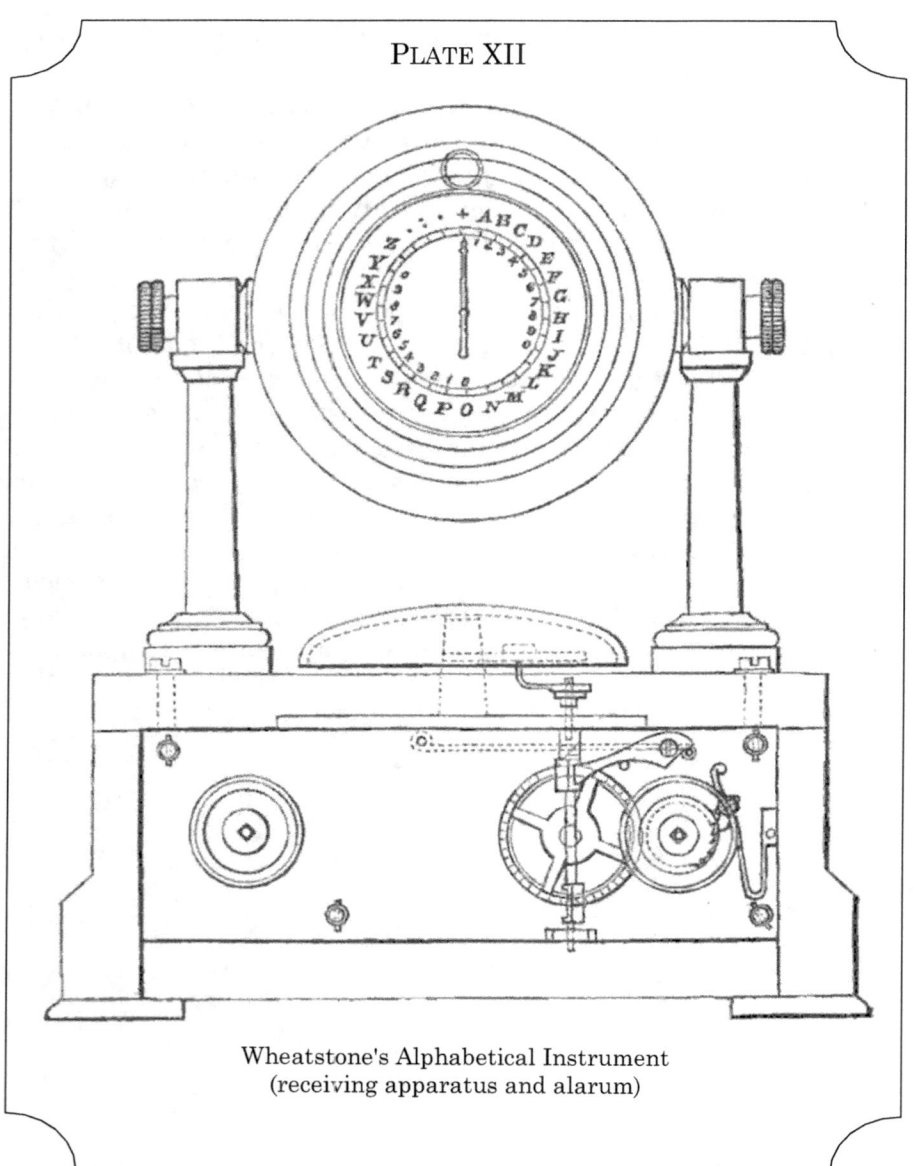

Wheatstone's Alphabetical Instrument
(receiving apparatus and alarum)

12
ALPHABETICAL TELEGRAPHS OR LETTER-INDICATING INSTRUMENTS

152. Professor Wheatstone's—153. Breguet's (French railway telegraph) —154. Siemens' (Prussian)—155. Lippens' (Belgian)—156. Froment's (French).

WHEATSTONE'S ALPHABETICAL INSTRUMENT

152. THE various step-by-step, or letter-indicating, telegraphs may be said mainly to owe their origin to Professor Wheatstone, whose early labours were mostly in this direction, and who has gradually perfected his apparatus until it has assumed the form now so extensively used for domestic purposes, and especially for communicating between offices, warehouses, &c. The advantage of this class of telegraph is, that those unaccustomed to its use can read the letters as they are successively indicated by the revolving pointer of the receiving apparatus; and, after a very short experience, can in like manner signal the letters with the sending keys.

From the nature of this form of instrument, and from the large number of currents required for each letter on an average, it is slower in operation than the needle, acoustic, or recording telegraphs. For this reason, combined with the difficulty of working it on any circuit of considerable length, the step-by-step indicating telegraphs have been pretty much restricted to the applications already mentioned.

In 1858, Professor Wheatstone patented a series of improvements upon the alphabetical instrument introduced by himself and Mr. Cooke in 1840. The object of the new apparatus was to attain greater steadiness, certainty, and rapidity than had been previously obtained where an index hand was moved rapidly, step by step, round an alphabetical dial. He embodied his ideas in a simple and yet portable form of instrument, which, from many advantages it possesses, has been very generally adopted for short lines where it is desirable to dispense with skilled telegraph clerks.

The currents used to work this apparatus are magneto-electric, evolved from permanent magnets; and thus the expense and trouble connected with the use of voltaic batteries in such an arrangement, are avoided.

The arrangement of this apparatus is shown in Plate XII, p. 198, and in Figures 85, 86, and 87.

Figure 85 is a vertical view of the outside of the sending apparatus, showing the small finger-studs *A A* (thirty in number) which, when pressed down, allow such a succession of currents to pass to the distant station as will move round the indicator on the receiving dial to the corresponding letter.

Figure 86 is a section (side view) of the sending apparatus. The handle *x* is used to turn a pair of electro-magnetic coils (*b b*)

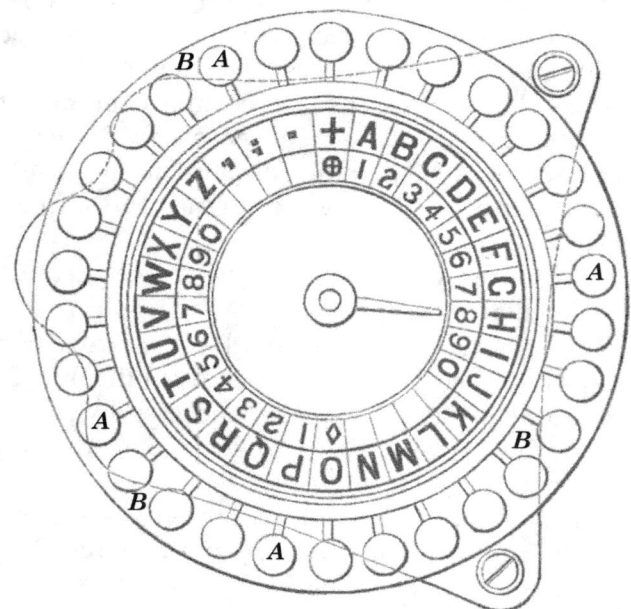

Fig. 85
Wheatstone's Alphabetical Instrument—sending apparatus

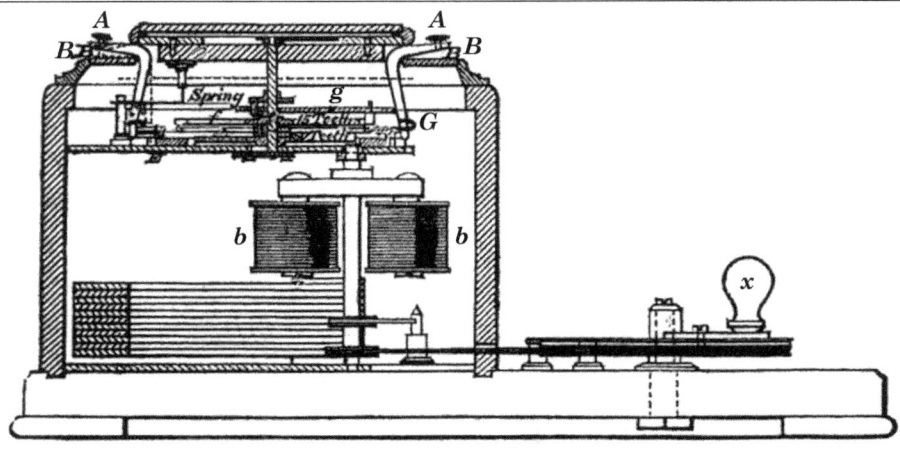

Fig. 86
Wheatstone's Alphabetical Instrument—sending apparatus (section)

by an endless band, rapidly above the poles of a permanent steel magnet *o*. A succession of magnetic currents are thus sent, and as long as the handle is turned, and the small finger-studs unmoved, the indicators of the receiving dials move on step by step. On one of the studs *A* being depressed to the stop *B*, the movement of the arm *g* is arrested as soon as it comes round to the lower end *G* of the finger-key, and the circuit is broken; no further effect being then produced on the indicating instrument, its pointer indicates the letter upon the dial to which attention is directed by the pause. An arrangement is made by which when a stud is depressed it remains down until another is pressed, upon which the first is mechanically returned to its former position.

Plate XII, p. 198 and Figure 87 show the indicating apparatus and alarum. The index, or pointer, is moved round by a pallet wheel acted upon by the alternate attraction and repulsion of two magnetic needles *h h* fixed upon an axis *g*, by the currents passed from the sending apparatus through the electro-magnets in the receiving instrument. *b c, d, e* form the bearings, and *h, i, j* is an external regulator.

The telegraphs which convey letters or words by conventional signs, like those described in the previous chapter, require a staff of agents engaged in their management, who have been specially

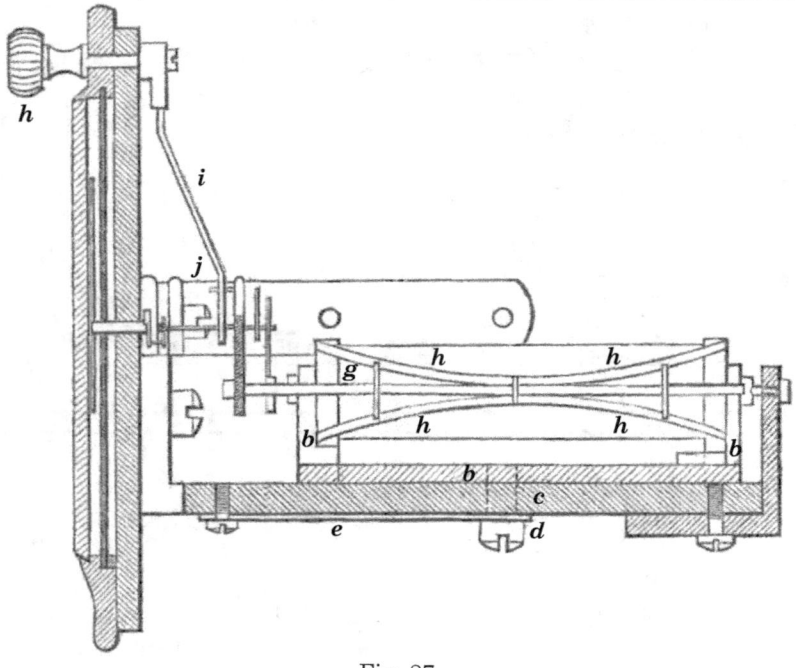

Fig. 87
Wheatstone's Alphabetical Instrument—indicating apparatus (section)

instructed and practised, as well in working the instruments as in interpreting their signs. As the railway telegraphs abroad at secondary stations are not in that constant requisition, and consequently do not occupy a permanent and exclusive class of agents, they are managed by any persons who happen to be employed in the respective offices: by the station-masters, clerks, railway police, guards, or, in short, by any railway agent who may happen to be at hand. Now, it is evident that telegraphic instruments, the use of which would require special instructions and much previous practice, could not be readily applied to such a purpose.

These considerations prevailed with the administrations of the lines of railway in many parts of the Continent, and led them at the outset to adopt telegraphic instruments which satisfied these conditions more completely than the apparatus adopted for state and public communications.

In general, therefore, the continental railway telegraphs are of the class of alphabetical instruments.

The French railway alphabetical instrument is in its principle identical with the state telegraph. The indicator in the latter, as described in the previous chapter, makes a complete revolution by eight successive steps, moving in each step through an angle of 45°. If the alphabet consisted of only eight letters, this would at once become an alphabetical telegraph by fixing the indicator in the centre of a dial upon which, at equal distances asunder, the eight letters are engraved. But since the French alphabet consists of 25 letters, and since an additional sign is found convenient, the dial is divided into 26 equal arcs instead of 8, and the indicator makes a complete revolution by 26 equal motions, at the termination of these motions respectively pointing to the letters engraved upon the dial.

To accomplish this, the escapement wheel is constructed with 13 teeth instead of 4, the groove upon the moveable disc of the commutator has 13 sinuous undulations instead of 4 sides with rounded corners, and the fixed disc upon which the handle of the commutator moves has 26 notches instead of 8.

The grooved disc, by the motion of which the oscillations right and left are imparted to the lever which makes and breaks the connection with the battery, is fixed immediately behind the notched disc, and the sinuous groove has the form represented in Figure 88 and acts upon the lever **G O H** by pressing the pin **G'** alternately right and left when the wheel **A B** is turned, the current being turned on and off by springs, **P, P'**.

The commutator, with its appendages, is represented in Figure 89. The fixed disc has at its edge 26 notches, into which the pin projecting from the handle falls, as in the state

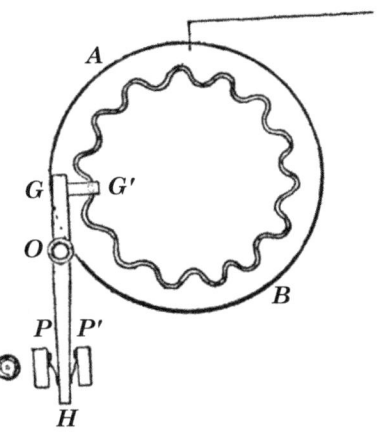

Fig. 88
Escapement wheel of
Breguet's telegraph

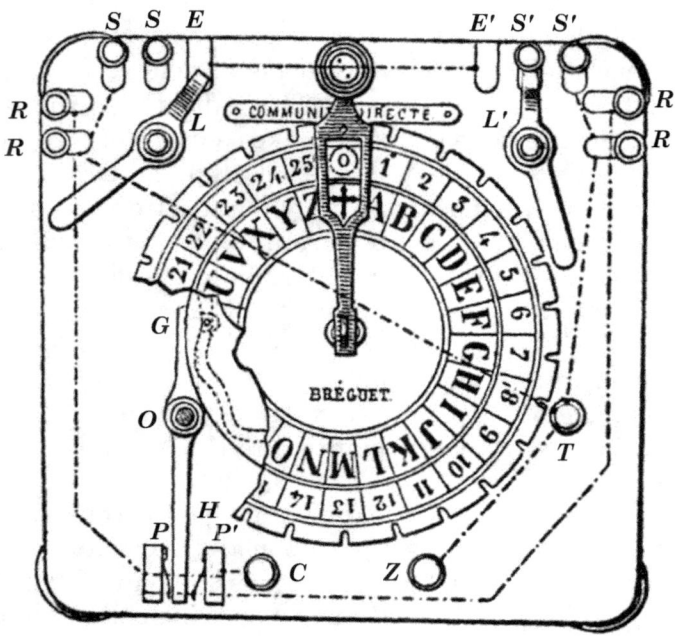

Fig. 89
Breguet's sending apparatus

telegraph. Engraved upon the face of the disc are, on the outside, the numbers from 0 to 25, and on the inside the 25 letters (W being omitted, not being generally used in the French language), the 26th place having the mark +.

A part of the dial is broken away, to disclose the face of the moveable disc, with the sinuous groove behind the fixed disc. The lever *G* is visible, with its pin in the groove, and the oscillation of the end of the lower arm *H* between the contact-pieces, *P* and *P'*, is exactly the same as that described in Figure 88.

The handle of the commutator is keyed upon an axis which, passing through the centre of the fixed dial, is itself keyed into the centre of the moveable grooved dial behind it, so that when the handle is carried round the fixed dial, the moveable dial behind is carried round with it.

Upon the upper part of the board carrying the dial are placed two supplementary commutators, *L* and *L'*, the hands of which play upon the contact-pieces, *S, S, E,* and *S', S', E',* as well as upon an oblong plate of metal, upon which the words "COMMUNICATION DIRECTE" are engraved.

The terminals *C* and *Z* communicate with the copper and zinc ends of the battery, or, what is the same, with its positive and negative poles; *T* communicates with the earth. The contact pieces *S S'* are connected with alarums, *R R* with the indicators, and the axes of the arms *L, L'* with the line-wires. The dotted lines indicate the positions of slips of metal inlaid in the back of the frame, by which the several pieces are put in metallic connection one with another.

After this general explanation of the manner in which the course of the current is in all cases governed, it will not be necessary here to explain the application of these commutating arrangements in other instruments of a similar character subsequently described.

A perspective view of the commutator and indicating apparatus mounted in the same case, is given in Figure 91. The commutator is fixed upon a horizontal desk, that being the most convenient position for its easy and rapid manipulation. The indicator, which corresponds with it in form, is placed like the dial of a clock in front of a vertical case.

If we suppose the commutator (Figure 89) at the station S, and the indicator at S', the arm of the commutator and that of the indicator being upon the mark +, any motion of the former made in the direction of the hand of a clock, will produce a corresponding motion of the hand of the latter, so that whatever letter or number the one points to, the other will at the same time indicate.

By this means the agent at S may spell word after word to the agent at S'.

There are various conventional signs, made by two or more complete turns of the handle of the commutator, which, being altogether arbitrary, and matters of local convenience, need not be noticed here.

It is found that moderately well-practised hands can transmit with this instrument seven or eight words per minute, while the most expert can send as many as eleven or twelve words.

A side view of the wheel-work and electro-magnet **E**, of the indicating apparatus, is given in Figure 90

The armature **P** is alternately attracted and dismissed by the magnet, acted on by the pulsations of the current, and imparts this motion to the escapement at **F**, by which the hand **A** of the indicator is advanced from letter to letter upon the dial, so that the motion of the hand **A** at the station S' shall correspond exactly with that of the hand of the commutator at the station S.

153. The telegraph which is represented in Figure 91, is a portable telegraph constructed for the French railways by M. Breguet. This instrument, in size and arrangement, is adapted to be carried in the guard's van upon the train, so that, in case of accident, it may be immediately put in connection with the line-wires, and notice of the circumstance may be instantly transmitted to the two stations between which the accident has taken place.

Portable instruments for a like purpose have been constructed in England and elsewhere.

The apparatus consists of a stout oaken case, containing in the lower part, **B B**, a Daniel's battery of eighteen pairs, a commutator, **M**, and an indicating apparatus, **R**. A small galvanometer is placed at **G**, to show the existence and force of the current, and a small electro-magnet, **L T**.

When not in use, the top, **C C**, attached by hinges to the case, can be turned down over the commutator and indicator, so as to close the entire apparatus.

A long rod of metal,

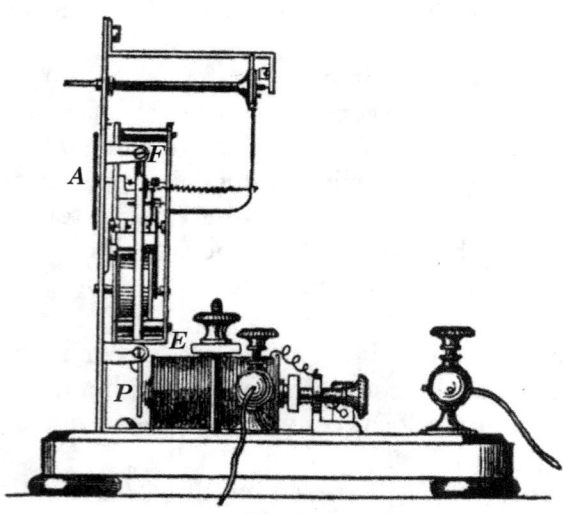

Fig. 90
Breguet's alphabetical instrument (section)

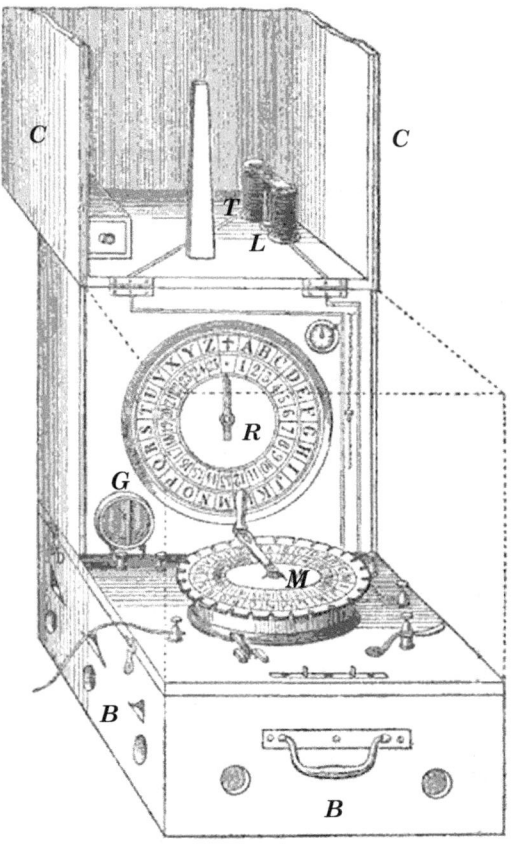

Fig. 91
Breguet's portable telegraph for French Railways

terminated in a copper hook, is provided, by which the end of the coil *L* can be put in connection with the line-wire; the end of the coil *T* being put in connection with the earth by means of a wire terminating in a small iron wedge, which is driven with a hammer into the joint between two of the rails.

To explain the manner of applying this apparatus, let us suppose an accident to happen between the stations S and S', and consequently the train to be stopped. The guard takes out the portable telegraph, and raising its cover, *C C*, he puts the wire of *L* in connection with the line-wire, and that of *T* within a joint of the rails, in the manner described above. He then makes one or two complete turns of the handle **M** of his commutator, observing whether the galvanometric needle **G** is deflected. If it is, he knows that he has transmitted a current to the line-wires. This current divides itself at the hook, and a part goes to each of the stations S and S', at each of which it rings the alarum. After a short interval a current is transmitted back from one or other of the stations, the arrival of which is indicated by the deflection of the galvanometric needle, **G**. The guard then informs the stations, one or both, of the accident, its place, the nature of the aid he requires, &c.

GERMAN RAILWAY TELEGRAPH.

154. The telegraphic apparatus used for the service of the Prussian railways, and for most of those of the German states, is one for which a patent was obtained by M. Siemens, of Berlin.

This apparatus consists of an indicating dial surrounded by the alphabet, upon which a hand moves similar in form and external appearance to the indicating dials of the alphabetical telegraphs already described, but placed upon a horizontal table, instead of being vertical as in the French telegraph. This dial is surrounded by a circular key-board, as shown in Figure 92, having as many keys like those of a pianoforte as there are characters upon the dial, the letter engraved upon each key being identical with that with which it corresponds in position upon the dial.

A lever, **a b**, is placed upon the table, turning upon the centre, **b**, and limited in its play by two stops, **T** and **R**. When it is turned against **T**, the line-wire is put in connection with the indicating apparatus, and when it is turned against **R**, that wire is put in connection with an alarum. A current, therefore, which is transmitted along the line-wire can be made to pass through the indicating apparatus or through the alarum at will, by giving to the lever **a b** the one position or the other.

The usual means are also provided by which the current may be allowed to pass the station without going through either the alarum or the indicating apparatus, or by which it may be stopped at the station and turned into the earth.

When no current passes upon the line-wire, and the instru-

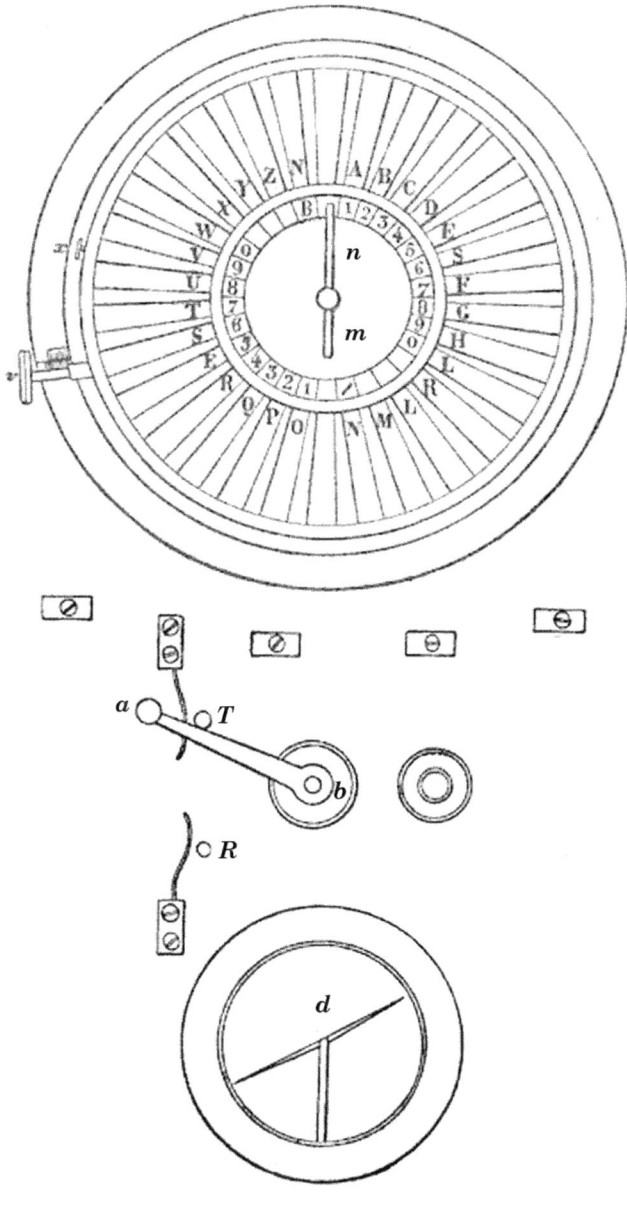

Fig. 92
Siemen's sending keys and indicator

ments are not in operation, the lever *a b* at each station along the line is placed against **R**, so that the line-wire is everywhere in connection with the alarum.

If it be desired to transmit a despatch from any station, S, the agent at that station puts the line-wire in connection with the poles of his battery, so that a current may be transmitted to all the stations upon the line. This current rings all the alarums, inasmuch as the arms *a b* are placed against **R** at all the stations. The agents at the stations being thus called, remove the arms *a b*, of their several instruments, and place them against the stops, *T*, the agent at the station S doing the same.

Previously to this, when the instruments were in repose, the indicating hands, *n*, in all of them were placed upon the division of the dial marked +. The moment the arms *a b*, or any of them, are placed against the stops *T*, the current transmitted upon the line-wire passing through the several indicating instruments, the indicating hands in all the instruments will commence simultaneously to move round the several dials. They will move from letter to letter with a starting and interrupted, but regular motion, like that of the seconds hand of a clock, but much more rapidly. The rate at which they are moved will depend on the force of the current; but, whatever be the rate, it will be common to all, all making successive revolutions of the dial precisely in the same time, and moving together from letter to letter with the most absolute simultaneity; and since they all started from the same point +, and move together from letter to letter, it follows that, whether their motion be quick or slow, they will all at each moment point to the same letter.

Now, it is important here to observe, that this common rotation of all the hands upon all the dials is produced and maintained by the current alone, without any manipulation whatever on the part of any agent at any station, and it would continue to be maintained indefinitely, provided that the battery were kept in action.

We have supposed the battery at the station S, from which the despatch is about to be transmitted, to be alone put into connection with the line-wire. But, in order to strengthen the current, each agent on the line, when he receives the signal, also puts his battery in like connection with the line-wire, so that the current acquires all the intensity which the combined action of all the batteries on the line is capable of producing.

The apparatus is so arranged that, in all cases, the galvanometer d is in connection with the line-wire, so as to indicate at all times at each station the state of the current.

It now remains to show how a despatch can be transmitted from any one station to all or any of the other stations on the line.

The apparatus is so constructed that if the agent at any station presses down any one of the keys surrounding the dial, the indicating needle, upon arriving at that key, will be stopped; and at the same moment the current upon the line-wire will be suspended. This suspension of the current will also, at the same moment, stop the motion of all the indicating hands upon all the dials on the line. The agents at all the stations will therefore see and note the letter on which the transmitting agent has put his finger. The transmitting agent, after a sufficient pause, transfers his finger to the key of the next letter he desires to transmit. The moment he raises his finger from the first key, the current is reestablished on the line-wire, and all the indicating hands rotate as before, passing again simultaneously from letter to letter until they arrive at the second letter upon which the transmitting agent has put his finger, when they again stop, and so on.

In this manner an agent at any station can stop the indicating needles at any or all the other stations successively, on their arrival at the letters of the words he desires to communicate.

If by reason of inattention or otherwise any letter or letters transmitted escape the attention of the agent at any of the stations to which the despatch is addressed, such agent immediately signifies the fact by putting his finger on one of the keys of his own instrument, by which he stops the hand upon the dial of the transmitting agent at a letter, which tells him to repeat the last letter or word, as the case may be. The signal is understood at all the other stations, so that no confusion ensues.

Having thus shown how a despatch is transmitted and understood by those to whom it is addressed, I shall now explain the mechanism by which these effects are produced.

Beneath the dial of each instrument an electro-magnet, such as *m m'* (Figure 93), is placed, upon the coil of which the current transmitted from the batteries passes. This magnet, then, as usual, attracts its armature *g o*, which comes against the stop *t'*. Now the apparatus is so arranged, that when *g* strikes *t'*, the circuit of the current is broken, and consequently the current is stopped. This deprives the electro-magnet *m m'* of its magnetism; and *g* being no longer attracted, it is drawn back from the stop *t'* by the spring *s*, and it recoils upon the stop *t*. Here the connection with the line-wire is reproduced, and the current is re-established. The electro-magnet having thus recovered its magnetism, *g* is again attracted by it, and drawn into contact with *t'*, where the connection is again broken, and *g* is drawn back to *t* by the spring *s*, and so on.

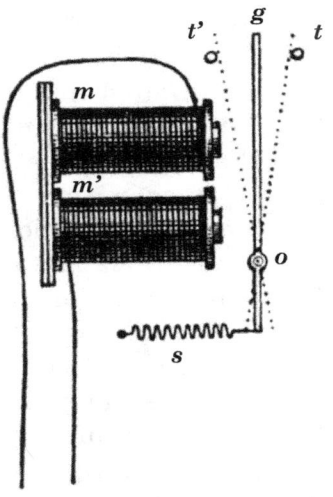

Fig. 93
Siemen's electro magnet

Since the intervals of transmission and suspension of the current are the same throughout the entire line, and since the intervals of transmission are those in which the armature moves towards the electro-magnet, and the intervals of suspension those in which it recoils from the magnet, it follows that the oscillations of the armature of all the electro-magnets at all the stations are absolutely alike and simultaneous.

In each instrument the armature is in connection with a toothed wheel, upon the axis of which the hand *m n* (Figure 92) is keyed, so that each vibration of the armature puts forward one tooth of the wheel, and advances the hand *n* from one letter to another.

Upon comparing this arrangement with that of the French telegraph, it will be perceived that here the mainspring and wheelwork which move the indicator are altogether omitted, and the armature of the electro-magnet, which in the French instrument only *regulates* the motion of the indicator, here both *moves* and *regulates* it. In fine, the armature here discharges at once the functions of the mainspring, and of the pendulum of a clock.

It will also be observed that the manipulation of the transmitting agent, by which he moves the indicators on the dials of the distant stations, is dispensed with, the current itself, through the intervention of the armature of the electro-magnet, imparting to the indicator a constant motion of rotation without any manipulation whatever.

That part only of the manipulation by which the indicator is stopped for a moment successively at the letters of the word intended to be transmitted, is retained, and that is effected by the action of the keys surrounding the dial.

Under the dial, a radius or arm is keyed upon the axis on which the indicating hand is fixed, so as to be always immediately under that hand and parallel to it, revolving simultaneously with it. This radius is a little longer than the indicating hand, and extends under the keys surrounding the dial. From the under-surface of each key a pin projects, the length of which is such that when the key is not pressed down, the radius passes freely under it; but when the key is pressed down, the pin comes in the way of the radius, and stops it when the indicating hand *n* arrives at the letter engraved on the key. By the action of the same pin the armature *o g* (Figure 93) of the electro-magnet is arrested in its return from *t'* to *t*, so as to be prevented from arriving at *t*. The current, therefore, is prevented from being re-established on the line-wire, as it would be if *g o* were permitted to come into contact with *t*.

Thus it will be understood how by putting down a key the two desired effects are produced. 1st, the stoppage of the indicating needles at the letter engraved on the key of the indicator on which such key is put down; and 2nd, the simultaneous suspension of the current along the entire telegraphic line, by which the indicating needles of all other instruments are stopped at the same letter.

This apparatus, compared with the French telegraph, to which it has an obvious analogy, has the advantage of greater simplicity. By dispensing with the mainspring and its necessary train of wheel-work, and with the rather complicated commutator worked by the hand of the transmitting agent, many moving parts are rejected, and there are proportionately less chances of derangement and less causes of wear or fracture. But on the other hand the moving power which impels the indicator, being transferred from the mainspring to the current, a proportionately greater force of current is necessary. This force is, however, obtained without augmenting the magnitude of the batteries at any one station by the expedient of bringing the piles of both the terminal stations, and, if necessary, of any or all the intermediate stations, into the circuit.

In the batteries used with the French railway telegraph, the use of acid, as has been stated, is found altogether unnecessary. In the German telegraph, however, pure water does not give a sufficiently strong current, and it is acidulated with about one and a half per cent, of sulphuric acid. The battery at each station consists usually of from 15 to 20 pairs. The usual speed imparted to the indicator by the current is about 30 revolutions per minute.

M. Siemens invented mechanism by which the indicating apparatus was connected with one by which the letters of the despatch as they arrived were printed by ordinary type upon a band of paper. Since, however, this has not been brought into practical use, it will not be necessary to explain it.

When the electric telegraph was first opened to the general service of the public in Prussia, this apparatus of Siemens was generally used, but it has since been superseded by that of Morse, its speed of transmission being found insufficient for the public service.

BELGIAN RAILWAY TELEGRAPH.

155. When the electric telegraph was first brought into use on the Belgian railways, the French and German apparatus described above were tried in succession. In 1851 they were, however, superseded by a form of telegraph invented and constructed by M. Lippens, mathematical instrument maker of Brussels.

M. Lippens attributes to the French and German railway telegraphs certain defects, which he claims to have removed, For the efficient performance of those telegraphs, it is evident that a certain relation must always be maintained between the force of the spring *s* (Figure 93), which produces the recoil of the armature *g o*, and the attractive force of the magnet, or what is the same, between the spring and the intensity of the current, with which the attraction of the magnet must vary. Now the intensity of the current is subject to variation, depending on the state of the battery, the number of pairs which are brought into operation, the length of the line-wire upon which it is transmitted, the more or less perfect state of the insulators, and in fine on the weather.

If the current become so feeble that the attraction of the magnet is less than the force of the spring *s*, the armature *g o* will remain upon the stop *t*, from which the magnet is too feeble to remove it. If, on the other hand, the spring have not sufficient force to overcome the friction and inertia of the armature *g o*, and the small portion of magnetism which may be retained by the electro-magnet after the current has been suspended, the armature will remain upon the stop *t'*, the spring being unable to produce its recoil.

Since, therefore, the forces against which the spring *s* acts, and which it ought to exceed, and those which act against it and which ought to exceed it, are variable, it is clear that the maintenance of the efficiency of the apparatus requires that the spring *s* shall from time to time be adjusted, so as to be kept in that relation to its antagonistic forces, which are necessary for the due performance of the telegraph.

It has been already shown that very sufficient and very simple means of adjustment for this purpose have been supplied in the French telegraphs. The hands which appear in the upper corners of the instrument (Figure 91) are intended for this purpose, and being turned by the key, the springs connected with them are increased or diminished in their force, according as the key applied to them is turned the one way or the other. Similar adjustments are provided in the German instruments.

M. Lippens and the Belgian railway and telegraph authorities by whom he has been supported, however contend, that although the permanent staff of the state and public telegraphs, constantly occupied and practised in the manipulation of such apparatus, may be relied upon for the due management of such adjustments; the agents of various grades employed on the railways, whose duties do not permanently connect them with the telegraph, and who are only called to it from time to time, cannot be depended on to perform adjustments requiring not only constant practice, but some address and some special knowledge of the principle and mechanism of the apparatus.

The apparatus of M. Lippens, which is now used for the service of the Belgian railways, is exempt from these defects.

Like M. Siemens, M. Lippens rejects the mainspring and its appendages adopted in the French telegraphs, and charges the current itself with their functions. He retains, however, the commutator, and imparts the pulsations to the current by the hand of the agent applied to a lever or winch, which is moved exactly like the arm of the commutator of the French instruments.

He rejects the spring *s* (Figure 93), which produces the recoil of the armature, and substitutes for it a second magnet placed on the other side of the armature, adopting at the same time a permanently magnetic bar of steel instead of the armature of soft iron used in the other instruments.

To explain the principle of Lippens' apparatus, let *a b* and *a' b'* (Figure 94) be two electro-magnets made precisely alike, the coil of covered wire upon them being one continuous wire carried from one to the other, and rolled in such a manner that their polarity shall always have contrary positions in which ever direction the current may be transmitted on the wire. Thus, if *a* be a north pole, *b'* opposed to it will be a south pole, and in that case *a'* will be a north and *b* a south pole. If the current upon the coil be re-

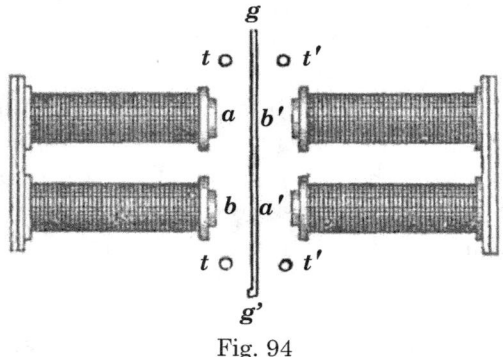

Fig. 94
Lippens' apparatus (part)

versed, all these four poles will at once change their names — *a* becoming a south, and *b'* a north pole, and *a'* a south and *b* a north pole.

Let *g g'* be a steel bar which is permanently magnetised, *g* being its north and *g'* its south pole, and let it be supported midway between the electro-magnets, having free play towards the one or the other until it encounters the stops *t t* or *t' t'* by which it is arrested.

Now let a current be transmitted upon the wire, by which *a* will become a north pole, and consequently *b* and *b'* will be south poles, and *a' a* north pole. Since *g* is a north and *g'* a south pole, they will be attracted by *b'* and *a'*, and repelled by *a* and *b*, and consequently the armature *g g'* will be moved towards *b' a'* until it is stopped by *t t*. If the current be then reversed, *a* and *a'* will become south, and *b* and *b'* north poles; and the armature will be attracted by *a* and *b*, and repelled by *b'* and *a'*, and will accordingly move towards the former until it is stopped by *t t*.

If the direction of the current be reversed rapidly, suppose, for example, ten times per second, the armature *g g'* will be made to oscillate ten times per second between the stops *t t* and *t' t'*.

It is evident that the expedient adopted by Siemens, by which the transmission of the current is arrested by the contact of the armature with one stop and re-established by its contact with the other, might be easily modified so as to reverse the direction of the current by each contact with *t t* and *t' t'*; and in that case the telegraph of Siemens would without other change be rendered exempt from the defects imputed to it, as well as the French instruments, by Lippens. But M. Lippens, either prevented from adopting this obvious expedient by the patent of Siemens, or giving a preference to the hand commutator for other reasons, has contrived an ingenious commutator worked by hand, by which he reverses the current with the greatest facility, rapidity, and precision.

This is a wheel commutator formed on the following principle:— There are two wheels placed side by side upon a common axle, with a disc of gutta-percha between them, so that one is insulated from the other. The edges of both are divided into a series of conducting and non-conducting arcs, but the position of these relatively to each other is alternate, the conducting arcs of each disc corresponding in position with the non-conducting arcs of the other.

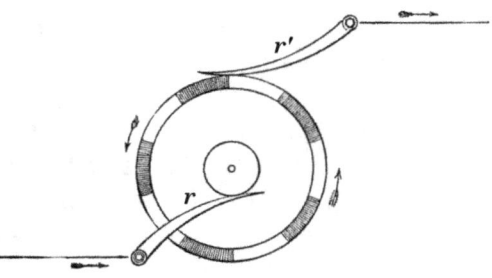

Fig. 95
Lippens' commutator

We may imagine the shaded arcs of Figure 95 to represent the conducting arcs of the upper, and the white arcs the conducting arcs of the lower disc, the one, however, being separated from all contact with the other by the interposed disc of gutta percha.

When the wheel is made to revolve, the spring *r'* comes alternately into contact with the conducting arcs of the one and of the other disc. Another similar spring is applied to another part of the edge of the wheel, so as to be in contact with the conducting arcs of the upper disc, while the spring *r* is in contact with those of the lower, and *vice versa*.

One of the two discs is in connection with the copper, and the other with the zinc end of the battery, so that one may be considered as its positive and the other as its negative pole. One of the springs is in connection with one end, and the other with the other end of the conducting wire, which forms the coils, and which passes along the telegraphic line. By causing the wheel to revolve, therefore, the conducting wire will be alternately connected with contrary poles of the battery, and the current upon it will be reversed.

If the edge of the wheel be divided into ten equal parts by the conducting arcs, this reversal will take place ten times in each revolution, and if a revolution be imparted to the wheel in each second, the current will be reversed ten times per second.

In the apparatus of Lippens the oscillations thus imparted to the armature *g g'* (Figure 94) are made to act by the intervention of toothed wheels upon the indicating hand which moves upon the dial around which the letters are engraved, as in the French tele-

Fig. 96
Lippens Belgian Railway Telegraph

graph, and this hand is moved from letter to letter in the same manner as in the French railway instrument and that of Siemens.

Upon the axle of the commutating wheel above described a winch is fixed, by which the agent who transmits the despatch turns it.

A plan of this instrument is drawn in Figure 96. The handle of the commutator ***B B'*** is keyed upon the axis of the wheel already described, which is under the table of the instrument. This wheel, and the springs which press upon it, are indicated in the figure. The handles **Q Q** are those by which the current is conducted from the up or down line through the indicating apparatus, or through the alarum, as already explained in the case of the German telegraph. Several other terminals are provided for establishing connections with the line wires, the battery poles, the alarums, and the earth, and differ in nothing essential from similar adjustments in other telegraphic instruments.

When the agent at any station, S, desires to transmit a despatch to any other station or stations, S', he first, as in other telegraphs, calls the attention of the agents at S' by means of the alarum. The current being then directed through the instruments severally by means of the adjustments provided for that purpose, the transmitting agent at S turns the handle **B B'** of his commutator, by which he produces the pulsations of the current, and puts the indicating hands upon the dials at S', as well as upon his own in motion. These hands as usual, when properly adjusted, always point to the same letters. The transmitting agent stops the handle **B B'** when he sees the hand **F** upon his dial point successively to the letters which spell the word he desires to transmit, and by continuing to operate thus, he transmits the entire despatch.

Such is the Belgian railway telegraph; and although it must be admitted that it supplies a certain improvement on the French telegraph, it ought also to be stated that the difficulty and inconvenience which M. Lippens claims to have removed, has not been found to offer any practical obstruction to the satisfactory performance of the French instruments.

FROMENT'S ALPHABETICAL TELEGRAPH

166. The external appearance of this instrument, represented in Figure 97, is that of a small pianoforte, having, however, no black keys. On each of the keys a letter of the alphabet is engraved, the first key being marked with a cross, and the last with an arrow. On the first ten keys are also engraved the numerals. This part of the apparatus is the commutator, by which the agent at the station where it is placed, is enabled to transmit signals to any distant

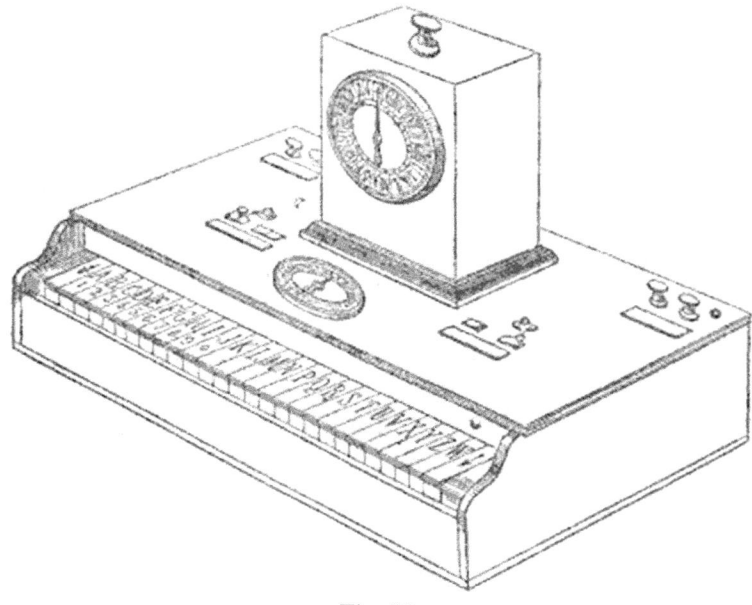

Fig. 97
Froment's alphabetical telegraph

station.

Upon it is placed the indicating apparatus, which is acted upon by the commutator of the apparatus at a distant station, and by which a despatch is received. This indicator is similar in form and in the manner of giving its signals to that of the other alphabetical instruments already described. The dial of the indicator is marked with the letters of the alphabet, and the cross and arrow corresponding with the characters engraved upon the keys of the commutators.

At the back of the case containing the indicating apparatus the alarum is attached, and commutators are placed upon the case by which this alarum can be put in connection at pleasure with the line-wire. As usual, it is always kept in connection with it when the instrument is not in use, so that notice may be given of the approaching arrival of a despatch. On the ringing of the alarum the agent at the station turns off the commutator from the alarum, and throws it into connection with the indicating apparatus.

To explain the transmission of a despatch, let us suppose an apparatus, such as that represented in the figure, to be erected at two stations, S and S', connected as usual by a conducting wire; the instrument, being unemployed, the line-wire at both is in connection with the alarum. Now let us suppose that S desires to transmit a despatch to S'. In that case S having first turned on the current, puts down any key whatever of his commutator, the effect of which is that a current is transmitted upon the line wire to S', which rings the alarum; then S' replies by transmitting a return current in the same way to S, by which S's alarum is rung. All being then prepared for the transmission of the despatch, S puts down with his fingers successively the keys of his commutator upon which the successive letters spelling the words of the despatch are engraved, and simultaneously with this the indicator upon the dial of S' points to the same letters, which are taken down by S'. At the end of each word, S puts down the key marked with the cross.

When it is intended to transmit numerals, S puts down the arrow just before he begins them, and the cross when he ends them. Thus if it be desired to transmit the number 1854, S first puts down the arrow, and then the keys marked A, H, E, and D successively, after which he again puts down the cross to indicate that the number is finished. It remains now to explain how these effects are produced.

Within the case, and at some distance below the keyboard, a steel rod is extended, parallel to the line of keys, the length of which corresponds with that of the row of keys. From this rod, and at right angles to it, proceeds a series of short steel arms, one under each key. In the bottom of each key, and at right angles to it, is inserted a short projecting pin, which corresponds precisely in position with the short steel arm just mentioned. The length of the arm, and that of the pin, taken together, is a little less than the distance between the bottom of the key and the steel rod when the key is not put down by the finger; the necessary consequence of which is that in that position of the key the rod may revolve, carrying the arm round with it unobstructed. But when the key is put down by the finger, the bottom of it is brought to a distance from the rod which is less than the sum of the lengths of the projecting arm and the pin; and consequently if the rod revolves, carrying with it the projecting arm while the key is thus held down, the pin coming in the way of the arm arrests it, and stops the further revolution of the steel rod.

It is evident that if the projecting arms were all inserted in the steel rod at the same side, or to speak with still more precision, if their points of insertion lay in a line along the side of the rod parallel to its axis, the pins of all the keys would arrest the revolution of the rod in exactly the same position; and as it will presently appear that the position in which the rod is stopped determines the signal transmitted, it would follow as a consequence that in such case all the keys would transmit the same signal, and the indicator at the station to which the despatch is to be transmitted would always return to the same letter upon the dial.

To prevent this, and to vary the signal in the necessary manner, the projecting arms are inserted in the steel rod according to a spiral or heliacal line, surrounding it like the thread of a screw; so that if, for example, the rod be placed so that the first projecting arm corresponding to the key marked with the cross, points directly upwards, the fourteenth which corresponds to the key M, will point directly downwards; and the intermediate arms will point at angles more and more inclined from the upward direction, each being deflected from the upward direction more than the preceding one, by the fourteenth part of the half circumference.

In like manner, in proceeding from the arm corresponding with the key M, which points downwards, each successive arm will be more and more deflected from the downward direction, each being more deflected from it than the preceding one by the fourteenth part of half the circumference.

Thus the twenty-eight projecting arms divide the circumference of the rod into twenty-eight equal parts; and consequently in a revolution of the rod, the arms come successively to the position in which they point upwards and in which they would encounter the pin projecting from the bottom of the key; if that pin were thrown in their way by the key being pressed down by the finger.

It will be evident, therefore, that if from any cause the steel rod be made to revolve, its motion may be stopped at twenty-eight different points of its complete revolution by means of the depression of the twenty-eight keys. We shall now show how a motion of revolution is imparted to this rod. To its right-hand extremity is fixed a ratchet-wheel, which is in connection with a train of clock-work, moved in the usual manner by a mainspring. This clock-work is contained within the case of the apparatus. If it be wound up, and if nothing obstructs its action, a motion of continuous rotation will be imparted to the ratchet-wheel, and by it to the steel rod; and this motion will be more or less rapid according to the force of the mainspring, and the adjustment of a fly which is connected with it. They are so adjusted as to cause the rod to revolve two or three times in a second. But in the teeth of the ratchet-wheel, a catch is inserted, which counteracts the mainspring and prevents the motion, which can only take place when this catch is withdrawn. A bar is suspended parallel to the keys, and under them, by a contrivance called in mechanics a parallel motion, by means of which any of the keys when pressed by the finger will lower it. This bar rests upon the arm of the catch engaged in the teeth of the ratchet-wheel, so that whenever any key is put down by the finger, the bar is depressed, the catch disengaged, the wheel liberated, and a motion of revolution imparted.

On the left hand extremity of the steel rod is fixed a commutating wheel, similar in principle to that already described in the railway telegraph. This wheel, being fixed upon the rod, turns with it, moving when it moves, and stopping when it stops. Since the position in which the rod stops is determined by the key put down, the position in which the wheel (thus fixed on the rod) stops is similarly determined. This wheel determines the pulsation of the current, and these pulsations determine the position of the indicator at the station to which the despatch is transmitted, in a manner which is substantially the same as those already described.

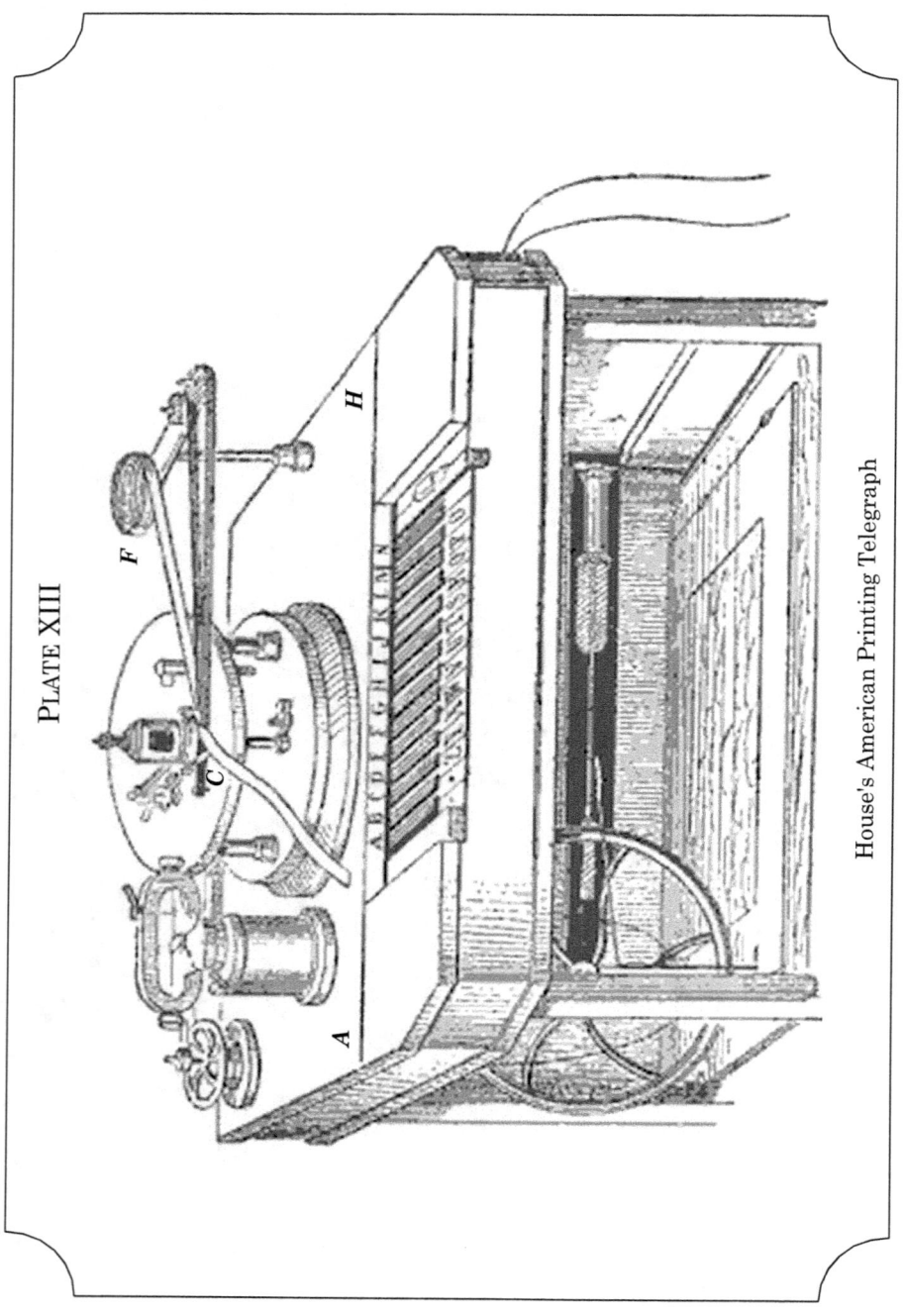

PLATE XIII

House's American Printing Telegraph

13
LETTER PRINTING TELEGRAPHS

157. House's American type printer—158. Brett's English type printer—159. Hughes' improved printing telegraph—160. Bonelli's chemical type instrument.

157. THE House telegraph, which was brought into use in the United States, is an example of the class of printing telegraphs, that is, instruments which print in the ordinary letters the despatch at the station to which it is addressed, by means of a power worked at the station from which it is transmitted. In a certain sense, this is accomplished by the various forms of recording telegraph already described, but in these cases the despatch is printed or written in cipher, which is attended with the inconvenience of being understood only by those who possess, and are sufficiently familiar with the key. The process of deciphering it, and writing it in common characters, occupying more or less time; for some purposes, such for example as that of journalism, this time must be taken into account in estimating the practical celerity of communications; inasmuch as the despatch, until so interpreted, is not available by the parties to whom it is addressed.

A telegraph which instead of impressing on paper characters in cipher, would impress the characters of common letter-press even though these should be transmitted and impressed at a slower rate than that of the transmission of the characters in cipher, might nevertheless be, in effect, more expeditious; more time being saved by superseding the process of interpreting the cipher, than is lost by the relative slowness of the transmission.

It is evident that these observations, being general, are applicable, not only to the instrument we are now about to describe, but to all others of the same class.

As the action of several trains of wheels is required in moving on the type wheel as well as the ribbon of paper, and in stopping both at the required instant when the impression of the letter is made; the intricate and expensive construction of mechanical type-printing telegraphs has hitherto prevented their general adoption. It is found that a specially skilled class of operators is necessary to work and regulate them.

House's printing telegraph, like all other telegraphic instruments, consists of two distinct parts: a commutating apparatus to govern the transmission of the current, and a printing apparatus upon which the current arriving from a distant station operates.

The manner in which the transmission of the current is controlled by the keys of the finger-board, is substantially the same as in Froment's telegraph. The wheel, however, that produces by its revolution the pulsations of the current, is moved, not as in Froment's by clockwork, but by the foot of the operator, acting upon a treadle like that of a lathe, which is seen under the case of the commutator in Plate XIII (p. 226).

The rotation of this wheel is arrested at the point corresponding to any desired letter, by putting down with the finger the key upon which that letter is engraved; in exactly the same manner and by the same mechanical expedient as in Froment's telegraph.

The keys, upon the keyboard of this instrument, govern by means of the pulsations of the current the motion and position of a dial or wheel at a distant station, inscribed with similar characters, in the same manner as has been already explained in the case of the French railway telegraphs.

Let us then suppose that by putting down any key, that inscribed with "A" for example at the station S, a certain dial or wheel at S', having upon it letters corresponding with those of the key-board at S, is so moved that the letter "A" is brought into a certain position. The letters upon this wheel are formed in relief like type, and when successively brought into the necessary position by the action of the current, having previously passed in contact with an inking apparatus, a band or ribbon of paper is pressed against them by means provided at the station S', and the impression of the letter is made upon the paper. By the next action of the current, the succeeding letter transmitted is brought to the same position, the ribbon of paper being meanwhile drawn forward, another impression takes place, and .so on.

The apparatus by which the ribbon of paper is moved, the type inked, and the paper pressed against it is not worked by the current. That process is effected by mechanism put in operation by the agent at the station at which the despatch is received and contained in the case *A H*.

In the figure, the ribbon of paper is represented at *F*, upon a roller from which it is gradually drawn, as letter by letter the words of the despatch are impressed upon it at *C*. The black band which appears upon another roller is an endless strap by which the types are inked.

In the mechanism as well of the transmitting as of the receiving apparatus, there are many details showing much ingenuity of contrivance, and resources of invention; which, however, are too complicated to admit of any clear exposition without numerous plans and sections, and which we must pass over.

The printing apparatus, at the station at which the despatch is received is put in operation by the action upon the treadle in the same manner as in the transmitting apparatus at the other station.

The galvanic apparatus, which supplies the current for working this apparatus, is the battery of Grove. About thirty cylindrical pairs are necessary for a distance of 100 miles. The first line operating with this apparatus was established between New York and Philadelphia in 1849.

HOUSE AND BRETT'S ENGLISH PRINTING TELEGRAPH.

158. The late Mr. Brett, who obtained celebrity by his successful exertions in establishing electric communication by submarine cables between the United Kingdom and the continent of Europe in 1850-1, took out, conjointly with Mr. House, a patent for a printing telegraph, the original form of which is represented in Figure 98.

The apparatus, like that of House's American telegraph, already described, consists of a keyboard, which is the transmitting apparatus

Fig. 98
House and Brett's English type printer

or commutator, and does not differ in any important particular from that already described. The receiving and pointing apparatus is also very similar, and stands upon the key-board. In front is an indicating dial, the hand upon which points successively to the letters printed upon the scroll of paper by the apparatus behind the dial. The printing apparatus, with modifications, is similar to that of House's American instrument.

159. Professor Hughes, of America, has invented a modification of the letter printing telegraph, which has considerable advantage over the previous instruments of this kind. Instead of sending a series of separate currents to bring round the letters of the alphabet consecutively, according to the number of currents sent; he uses a single current for each letter. The duration of the current determines the letter to be printed, by acting upon the rotating type wheel at the distant station.

Such a plan necessitates the train of wheels moving isochronously at both ends, and to produce this a short spring pendulum

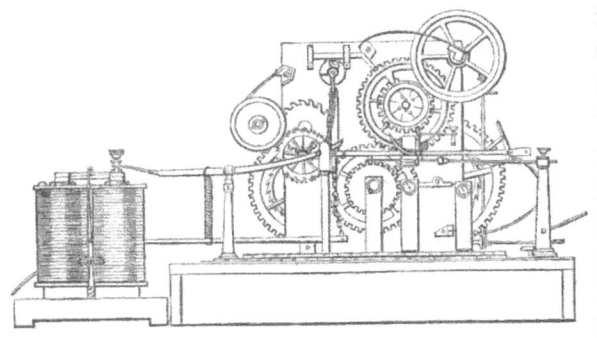

Fig. 99
Hughes' American printing telegraph

is employed to regulate the clockwork escapement of the apparatus. The vibrations of the pendulum can be varied in speed by shifting a weight. This instrument is shown in Figures 99 and 100. It is now used to some, extent in America, France, England, and elsewhere.

160. Mr. Bonelli has also introduced a form of letter-printing telegraph, which is a modification of the Bain chemical or Bakewell copying telegraph. The instrument requires five conducting wires, each in connection with a platinum point resting upon part of a letter set in type. By moving the type frame forward a series of

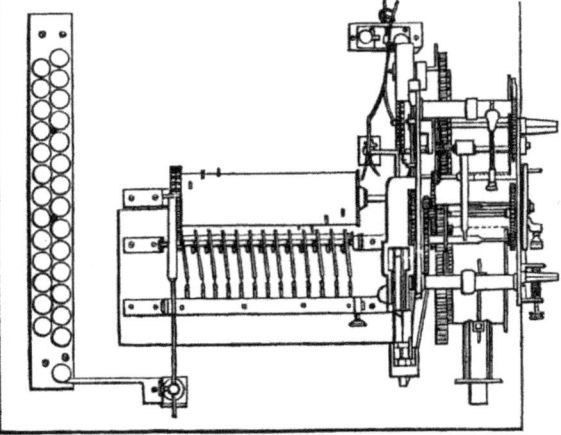

Fig. 100
Hughes' printing instrument (vertical plan)

contacts are made as the points press upon the different projections of the letters, and corresponding marks are thus made by the currents passing through a riband of chemically prepared paper at the distant station.

Of course the practical objection to this form of instrument consists in the number of telegraph wires employed, and the time lost in setting up messages in type.

PLATE XIV

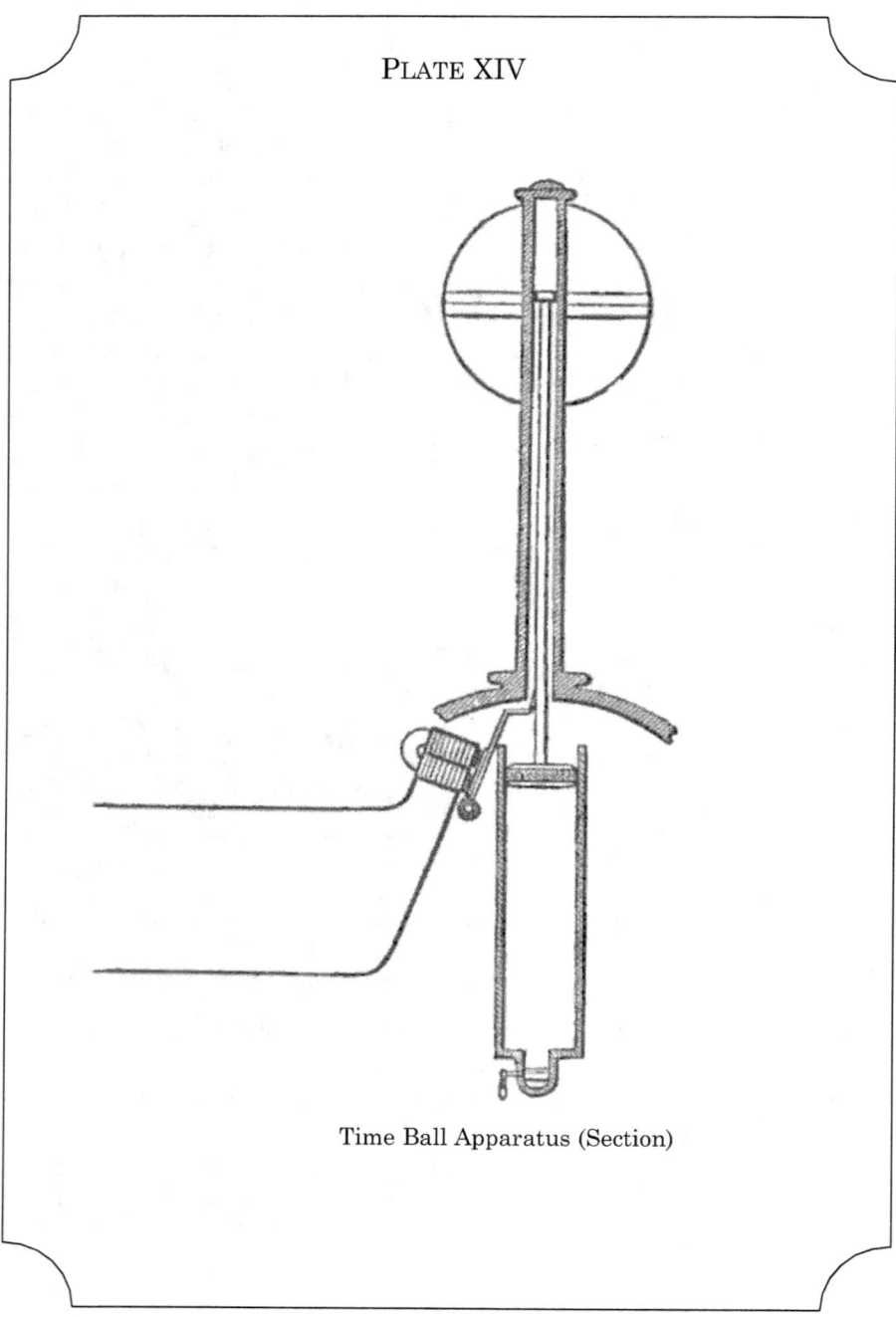

Time Ball Apparatus (Section)

14
THE TRANSMISSION OF TIME, WEATHER OBSERVATIONS AND STORM WARNINGS

161. The object of telegraphing accurate time—162. Electric time-balls—163. The regulation of clocks by electricity; objections to early methods—164. Improved plan of R. L. Jones, and its application in London, Liverpool, Glasgow, Edinburgh, &c.—165. Meteorological messages from telegraph stations to Board of Trade—166. Warnings telegraphed on indications of storms—167. Forecasts of weather.

161. THE telegraphic transmission of accurate time by electric currents from Observatories to public clocks and time-balls, has within the last few years become of considerable importance.

It may be said, "but what is the great benefit arising from extreme accuracy? If a clock keeps time within a few seconds, or even a minute or two, it answers the purpose." So it may for those who have no occasion to cross the sea; but to sailors exactitude in time is everything, and unless their chronometers are carefully rated and timed before they set sail, their errors may set a ship many miles out of its proper course, and lead to wreck.

162. Hence it is that for some years past time-balls have been regularly dropped at a fixed hour in London, Liverpool, Glasgow, Deal, &c., by currents from the nearest Observatory; thus enabling captains to check the going of their chronometers daily while in port.

The apparatus employed is very simple. The time-ball is retained at the top of a staff passing through its centre, by means of a catch. The soft iron armature of an electro-magnet forms one end of a lever connected with the catch. As soon as the electric current is passed from the Observatory the electro magnet attracts the armature, thus drawing away the catch and allowing the time-ball to drop by its own gravity. To break the force of its fall it is made to press down the piston of a partially closed air cylinder in its descent; the compression of the air acts as an elastic buffer, and gently arrests the lower part of its fall. A sectional view of this arrangement is shown in Plate XIV (p. 232).

The current is sent at an arranged moment by a clock at the Observatory, which is of course continually regulated to true time by observation. A glass or ivory disc, revolving on the same axis as the seconds wheel, has a small metallic contact at one point of its circumference, and this is brought against a spring at a given second, thus connecting up a battery with the wire leading to the time-ball.

163. Let us now see how public clocks can be regulated and controlled second by second from an Observatory.

The early methods of applying electricity to clocks failed, because too much was attempted; the currents were either made actually to drive the clock, or to release its escapement; and if anything failed in connexion with the battery, conducting wire, or other apparatus, the clocks were stopped. In addition to this objection the current required was so excessive that expensive forms of battery were necessary.

164. Now, however, by an invention of Mr. R. L. Jones, the current is simply made to act as a sort of bridle or check upon the motion of the pendulum bobs of an ordinary clock, the only altera- tion consisting in an electric coil of equal weight being substituted for the metallic bob. Two small steel magnets are fixed, one on each side of the centre of the arc of vibration, so that the coil bob moves alternately over one or the other during its beat.

A regulator clock at the Observatory is so arranged that its pendulum at each end of its beat presses a light spring and makes contact with a battery, thus sending a short current each second through the wire coil forming the bob of each public or other clock, to which the conducting wire from the Observatory is joined. When the current passes, the coil pendulum bob of each regulated clock becomes momentarily magnetised; and its poles, thus created, exercise a mutual action, upon the poles of the steel magnet over which it passes.

This method of regulation is shown in Figure 101. *A* is the pendulum of the regulator at the Observatory, in connexion with

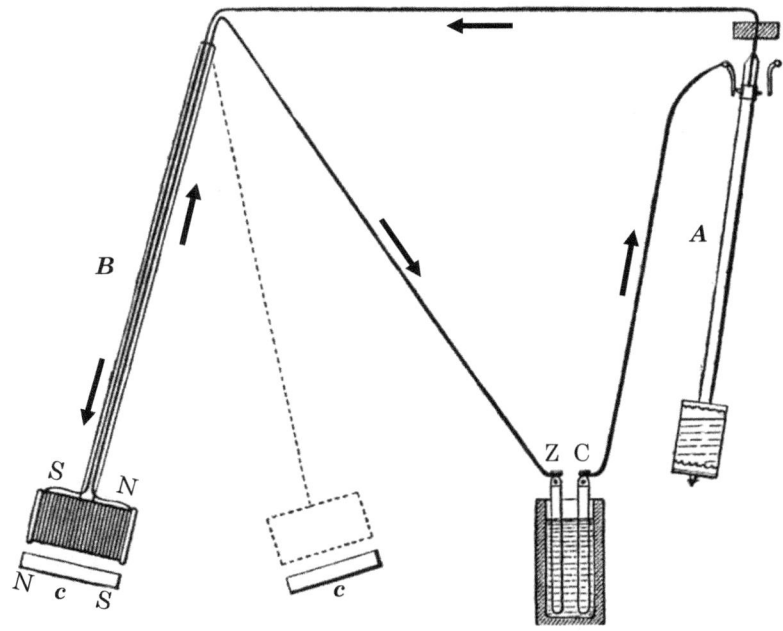

Fig. 101
Diagram of Jones' method of clock regulation by electricity

the battery and conducting wire; **B** the pendulum of one of the public clocks regulated, with its bob formed by a wire coil. The fixed steel magnets, **c c**, over which the bob of **B** passes in its beat, are shown below the coil, but in practice are so placed that they enter the hollow centre of the pendulum coil as it moves.

If the pendulum is coincident in beat (when polarised) with the Observatory pendulum, and therefore at the required point of its vibration when the current arrives, its motion will not be affected by the magnetic action between the coil bob and the fixed magnet. But if on the contrary the normal, or uncontrolled, rate of the regulated clock is either fast or slow; in the former case a slight retarding, and in the latter an accelerating, impulse will be communicated to the pendulum by the corrective action of the electric current. The effect of this is either to check or pull on the pendulum during each of its vibrations.

As a clock, even with a great rate of error, cannot vary much in the course of a second, it is found in practice that large public clocks may be brought under the most absolute control by the influence of an exceedingly feeble battery. The system was first applied about ten years ago to the control of various public clocks at Liverpool by Mr. Hartnup, the astronomer, in conjunction with the writer; and has since continued to afford the most satisfactory results. A seconds clock on the Exchange has enabled the various clock-makers to regulate their chronometers with the utmost exactitude; and the large Town-Hall clock with other public clocks have been kept truly to Greenwich mean time, with little or no trouble, and at a trifling cost even compared with the ordinary clockmaker's charge for regulation. The method has since been introduced in London, and has also been very extensively adopted in Glasgow by Professor Grant, and at Edinburgh by Professor Piazzi Smyth, in connexion with their Observatories. Some of the church and other public clocks regulated are old and ill-made, previously keeping wretchedly bad time — frequently minutes wrong; but since this simple electric check was applied, they have gone with the closest accuracy.

In the event of any temporary interruption, from accident or otherwise, to the transmission of the electric current, this system has the great advantage of not stopping the public clocks. They continue to go with their normal rate of error whatever it may be, until the conducting wire or battery is put right and communication with the Observatory regulator restored. It is needless to point out that the plan might be readily adopted in connection with the great clock in the tower of Westminster Palace, and with the various minor timepieces in the Houses of Parliament, so as to keep them all isochronous for the information of members.

A slight modification of the arrangement has been introduced by Sir Wm. Thompson in connection with the Glasgow clocks, by means of which "two seconds" pendulums can be controlled in conjunction with "seconds" pendulums on the same wire.

In Liverpool an addition is made by which each clock regulated sends a signal back to the Observatory, deflecting a galvanometer needle at a given second in each minute, as if to say "I'm right!"

165. The system of meteorological telegrams and storm warnings, devised some years ago by the late Admiral Fitzroy, and carried out by him under the auspices of the Board of Trade, has proved of great service to mariners; especially to those engaged in the coasting trade and fisheries, where the small size of the vessels employed, coupled with their sailing as a rule within a short distance of land, render them peculiarly liable to disaster from storms.

The arrangements are to collect, several times daily, simultaneous readings of barometers and thermometers, together with details as to the weather, from a number of points extending over an area of a thousand or more miles square; stretching between L'Orient and Rochefort in the south, Galway, Valentia, and Cape Clear in the west, and Nairn, Holder, and Skuddesnæss in the north and north-east. The meteorological office in London thus becomes aware of the approach of dangerous winds at the confines of the signalling area, a day or two before they reach our channels and ports.

Storms are for the most part circular, rotating upon a centre like an eddy in a river; and while the currents of air and vapour may be sweeping along very rapidly in a curve within the limits of disturbance, the body of the storm usually progresses comparatively slowly; so that ample time for warning exists, as a rule, after the indications of its approach are received from the outside observing stations.

In the temperate zone of the earth in which the British Islands are placed, the atmosphere has a general progression as a rule towards the east. Storms, or vast eddies between great breadths of antagonistic currents of air, are therefore translated eastward, being more or less deflected by local conditions towards the north or south; and having passed the meridian of any place are gone. Hence it is obvious that in temperate zones reporting stations are required north, west, and south of the central warning office; but not towards the east.

The observations are obtained in the following way:– Each of a number of signalling telegraph stations round the coast is supplied with a standard barometer, a wind vane, and two thermometers — one with a wet bulb. At 8 a.m. each stationmaster observes the index of the barometer and thermometers; the direction of the wind, the force of which he also estimates; the state of weather as to cloud, rain, snow, &c.; and also the amount of sea disturbance. He embodies all these facts in a concise form of cipher message, in which the results are expressed by a series of six groups of five figures each.

The following is a specimen message:–

GALWAY TO BOARD OF TRADE, LONDON.

29046 —
54513 —
50308 —
06534 —
05063 —
03322 —

EXPLANATION OF ABOVE

29.046	Height of barometer, in inches and decimals.
54	Height of thermometer, in degrees.
51.3	Height of exposed thermometer, in degrees and tenths.
50. 3	Height of moistened thermometer, in degrees and tenths.
08	Direction of wind in a series of 32 points.
06	Force of wind, estimated in 12 degrees.
5	Amount of cloud, estimated in 9 degrees.
3	Weather: Character of weather given by 9 figures.
4	Sea disturbance: amount given by 9 figures.
05	Average direction of wind.
06	Force of wind.
3	Weather.
033	Highest or lowest reading of barometer.
22	Highest or lowest reading of exposed thermometer.

} Since last report.

These messages are received from the various points by the meteorological office in London between 8 a.m. and 9 a.m.; a similar series of observations being also telegraphed at 1 p.m. Thus the staff at the central office have prompt data supplied on which to base their calculations; and are presented with a sort of bird's-eye view of the atmospheric condition of a great area, including the British Isles, south-west coast of France, and Holland.

166. Upon the advent of foul weather being thus indicated, the London office at once dispatches messages giving warning to

Cone up.
Northerly gale
expected.

Cone down.
Southerly gale
expected.

Drum.
Variable gales
expected.

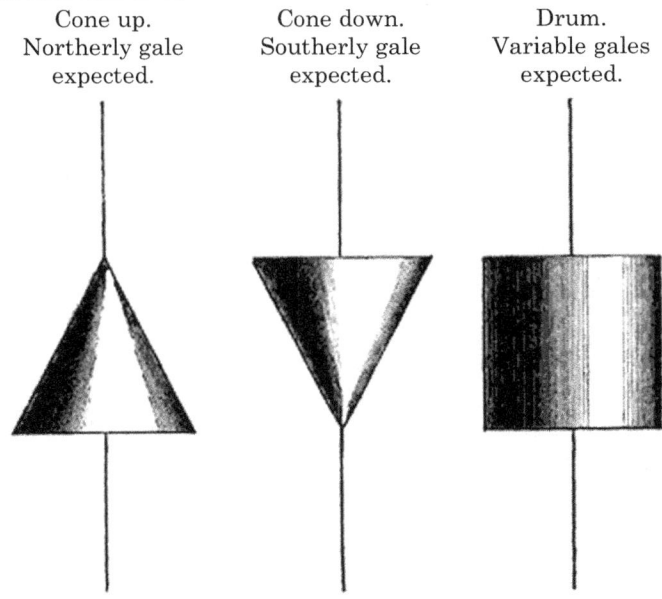

Fig. 102
Board of Trade storm warnings

the principal ports and coast-guard stations; and signals are hoisted without delay at conspicuous points, to give notice of the expected storm to vessels in the vicinity.

The storm signals hoisted to warn ships consist of a cone and drum, which are used to indicate the nature and probable direction of the expected gale. They are made of black painted canvas mounted on hoops. The cone when seen from any direction appears as a triangle, and the drum as a square. When the point of the cone is upwards, it signifies that the storm will probably commence from the northward; and if the point is down, from the south. The drum indicates dangerous winds expected in various directions successively. These signals are shown in Figure 102.

At night the signals are given by means of three lamps in a frame, to represent the cone or triangle; and four, for the drum or square. The objection has been raised that mariners may be made unnecessarily fearful of venturing to sea, by these warnings of an approaching storm, but surely this is not probable. All captains consult their ship barometers to judge of coming weather, as well as they can, in their own immediate vicinity; and how much better is it to be furnished at the same time with the condensed result of a vast series of observations over a greatly extended area. Forewarned on board ship is forearmed; and if a captain sails knowing the approach of bad weather and the quarter whence it is coming, he can get well clear of land betimes, and have everything ready on board to meet the storm.

167. The occasional storm warnings were for a time supplemented by daily forecasts of the weather from the meteorological office; but these have not as yet been found thoroughly reliable, although of course the inference drawn from the admonitions of a barometer are much more valuable when corroborated by simultaneous readings of many instruments spread over a continent. The further the telegraph is extended the longer will be the possible period of fore-casting; and now that America is linked by the wires to Europe, Asia, and Africa, it is probable that soon any important pulsations of our atmospheric ocean will be signalled days beforehand all over the globe, and a vast meteorological system carried on by telegraph. To exemplify the benefit derivable from Storm Signals and their accuracy, the following short extract is added from a Report of the Meteorological Department of the Board of Trade.

Date.	Caution transmitted.	Weather following Cautions.
1864. 9th Jan. 1 P.M.	South cone (western coasts).	A gale from S.S.E. to S.S.W. on the 11th, on the west coasts of England and Ireland.
11th Jan. 5 P.M.	Drum (on all coasts).	On the 12th the wind blew generally from S.E. to S.W., strong to a gale, but moderated as it changed to N.W. and N.E. The gale was more severe on the western and southern than on the eastern coasts.
16th Jan. 5 P.M.	South cone (on southern and western coasts).	A strong southerly gale continued from the night of the 16th till the evening of the 17th on the western and southern coasts ; and was felt as a strong wind on the eastern coasts on the 17th. *
21st Jan. 11 A.M.	South cone (on all coasts).	On the 22nd a strong south-westerly gale prevailed (heaviest on the south and west coasts), followed by strong westerly winds on the 23rd. The *Lucinde* of Lynn encountered this gale about 6 miles S.E. of Coquet Island and was disabled ; the captain speaks of the storm as a hurricane.
1st Feb. 1 P.M.	South cone (on all coasts).	A strong south-westerly gale on all coasts from the 2nd to the 3rd. The gale decreased as the wind

* The *Craigs* reports :—"On January 13th, lat. 48° N., long. 34 W., encountered a gale from S.S.W. At 8 A.M. wind shifted westerly, and at 10 A.M. came N.W. in a terrific squall ; the sails being immediately blown away in ribbons, such was the violence of the gale."—*Shipping Gazette.*

"The brig *Ianthe* from New York reports having experienced a succession of southerly and westerly gales the entire passage, and that on the night of the 13th, in lat. 49° 10′ N., long. 24° 20′ W., encountered a hurricane, commencing from S.S.W. round to W.N.W. At midnight, while lying to under close-reefed maintopsail and reefed stormstaysail, a heavy squall blew them to shreds, throwing the ship on her beam ends and shifting the cargo."—*Shipping Gazette.*

Date.	Caution transmitted.	Weather ollowing Cautions.
		drew round to W. and N.W. The weather was squally and unsettled, in the *north*, previous to the receipt of the caution ; but there was subsequently a *gale* from S.W. On some parts of the east coast the gale was severely felt; from Staithes it is reported that it blew a hurricane at that place throughout the night of the 2nd, with showers of hail and rain.
8th Feb. 1 P.M.	Drum (on northern, western, and southern coasts).	On the 10th a gale on the south and west coasts (the direction of the wind is variously given from all points between N.E. and S.) and strong north-easterly winds on the north coast. Snow fell very generally all over the country between the 9th and 11th. At Wick, "a gale raged from the night of the 9th to the morning of the 10th, doing serious damage to the fleet of fishing smacks assembled there ; one of which was lost with all hands."—*Shipping Gazette* and *John O'Groat Journal.*
10th Feb. 4 P.M.	Drum (on eastern coasts).	On the 12th and 13th a gale on the eastern coast ; the wind at first blew from between S.S.E. and S.W. and then changed to N.N.W. and N.W.†
12th Feb. 11 A.M.	South cone (on northern and western coasts).	Strong gale from S.W. to W. on the north and west coasts on the 13th.‡

† Grimsby, Feb. 14.—"Yesterday it blew a heavy gale from S.W. About 2.30 P.M. the wind suddenly flew to the W.N.W., blowing a complete hurricane with heavy rain."—*Shipping Gazette.*

Filey, Feb. 14.—"Yesterday afternoon a most violent storm of wind and rain came on suddenly, making the sea almost instantaneously one complete drift :—Observed at about a distance of four miles a large brig and a small schooner in company. Just at this moment a terrific gust struck the schooner and laid her on her beam ends : in another instant she appeared bottom up, and the next, every vestige seemed to disappear."—*Shipping Gazette.*

‡ Falmouth, Feb. 15th.—"Capt. John Swain, of the brig *Peep o' Day*, from Pembroke for Nantes. reports :—On the 13th bore up, in a heavy gale from S.W. from lat. 49° N., long. 5° 50' W., and arrived in Falmouth on Sunday (13th) at 1 P.M. Made another attempt to get to the southward on the night of the 13th, but found it impossible to do so on account of the heavy sea running in the channel."—*Shipping Gazette.*

PLATE XV

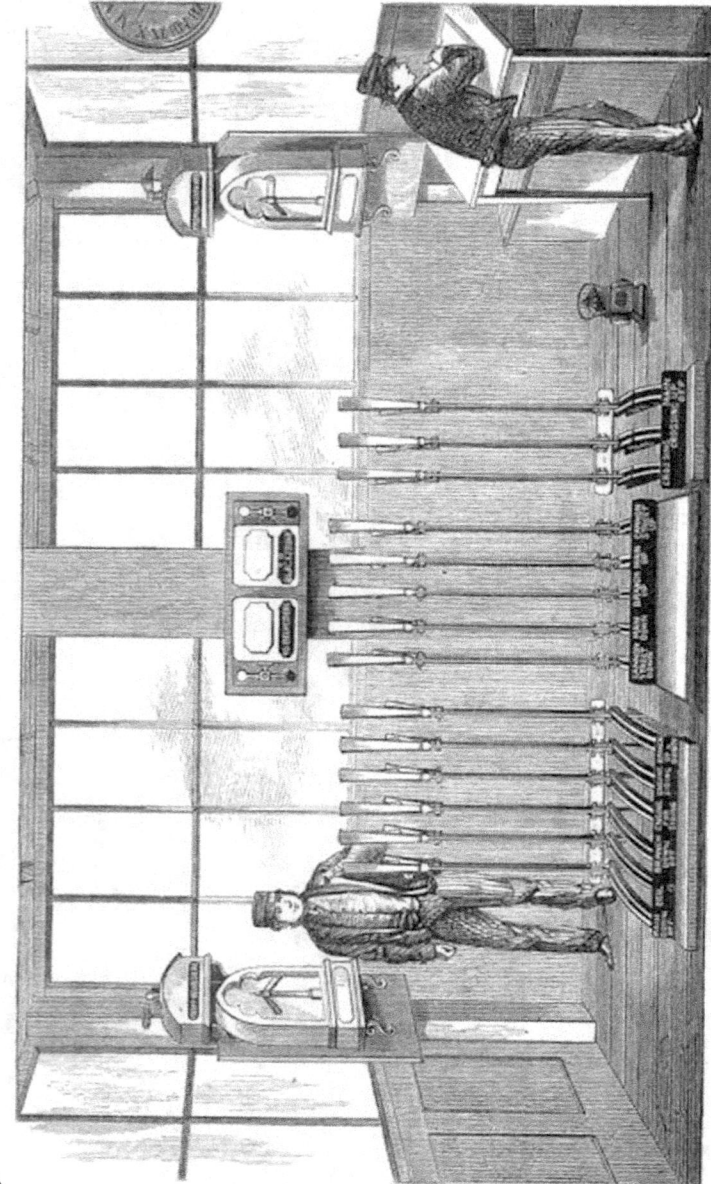

Signal-box at Waterloo on the Charing Cross line

15

RAILWAY TELEGRAPHS
AND TRAIN SIGNALLING

168. The electric telegraph essential to railways—169. Their use of conversing apparatus—170. Principles of train signalling—171. Messrs. Walker's telegraph semaphore—172. Preece's electric semaphore—173. Tyer's train signal apparatus—174. Disc railway telegraphs of Messrs. Bartholomew and Spagnoletti—175. Communication between passengers and guard.

168. Railways were the first to give practical application to the telegraph in this country; and it has most amply repaid them. Its use in connection with the working details of the railway network has continually extended, until now the advantage of electric agency in carrying on these vast undertakings can scarcely be over-estimated. While some wires appropriated to the conversing instruments are constantly employed in giving directions connected with the management of the concern, as to engines, carriages, passengers, goods, luggage, &c; other wires even yet more important in their use are silently engaged in signalling on the passing trains from section to section, and guarding them from injury. By other apparatus, in case of a train breaking down from any cause the guard is enabled to signal right and left to the stations on each side, by means of which following trains are stopped and assistance summoned. The use of the telegraph on railways does not stop here, but is now introduced into the trains themselves, so that the guard and driver may communicate, and passengers, in case of need, may summon the guard to any compartment of the train by a simple signal.

The railway conversing telegraph is, of course, one form or other of the ordinary instruments used by the telegraph companies, and does not therefore need particular description; but it is found, as a rule, that the less complicated the apparatus is, the better, as it has to be worked and looked after in many cases by the porters and others in the railway employ who have little or no knowledge of electrical arrangements; usually a form of needle telegraph is employed for this purpose.

169. Before proceeding with the description of train signalling apparatus, and with that used for communication between passenger and guard, the various employments to which the conversing telegraph is put may be briefly summarised. Rolling stock and stores are telegraphed for, and promptly sent from one point to another whenever wanted. The numbers of passengers of each class coming forward from other railways, or about to arrive at junctions, are accurately transmitted, enabling the requisite carriages to be provided, where without such information there might be too few, or far too many kept idle at the junction station. In the same way, livestock and cargo shipped from various ports are signalled to the railway termini for which they are destined, and at which onward carriage will be required, and thus the requisite wagons are ready in advance of the arrival of the vessel.

In fact, wherever a demand for extra conveyance arises, the conversing telegraph steps in to aid, and the sensitive wire fibres convey the intelligence to headquarters or the nearest depot; and whether engines, carriages, wagons, or tarpaulins are wanted, they can be at once directed to the place in need.

If luggage is lost, a few movements of the needle in various directions track and restore it. Again, the use of the wires is of the utmost advantage in correcting errors in the transmission of goods and in expediting their delivery, by the freedom and promptitude with which enquiries on such subjects can be made and answered.

The two principal results of the railway telegraph are thus economy and safety — economy in rolling stock and stores, safety to passengers and staff; besides which, a far greater number of trains can be safely passed than the companies dare put upon their rails, especially on crowded sections, without this wonderful adjunct. The quick succession and varying speed of trains, coupled with the complication of points and approaches to the principal stations and junctions, have rendered essential this instantaneous and careful electric monitor, that might in the old times of slow and occasional trains perhaps be dispensed with.

Railway shareholders may congratulate themselves upon this practical realisation of the old fable of the mouse and the lion; many meshes encumbering our iron roads have been broken by the operation of the puny wires, and the profits from such undertakings immeasurably increased. It is indeed a question whether some railways would pay any dividends at all without the assistance of the telegraph in conducting their operations; and it is certain that without this agency on some of the more important lines, the rails along their busier sections would have to be doubled and in some cases quadrupled in number, at a ruinous outlay, in order to accommodate even the existing traffic.

170. In train-signalling it is customary to place stations within very short distances of each other near the termini or important junctions, and at greater distances on other parts of the railway, pretty much in proportion to the traffic passed; the principle being to prevent any train passing upon a given section between the two signalling points, until the preceding train has been signalled forward from the station in advance. With this precaution put in force, no two trains can be upon the same section at once, and the chance of collision is thus precluded. This plan is known as the "block system," and is of course somewhat costly to introduce; but when we look at the expense to the railway resulting from a single accident, apart from the far more important object of saving life, the outlay upon such a mode of safeguard sinks into insignificance.

Before describing the various plans in operation, some principles may be referred to that should be characteristics of any train signalling arrangement. The first essential is "current equilibrium" — the batteries in the stations at the opposite ends of each section being connected by their similar poles through their respective signalling apparatus to the wire, as shown in Figure 103 overleaf.

The currents thus tending to pass from each battery, are opposed and balanced. On the balance being disturbed by anything being wrong with battery or wire, the instruments are at once thrown out of order, and signals given at *both* ends. The

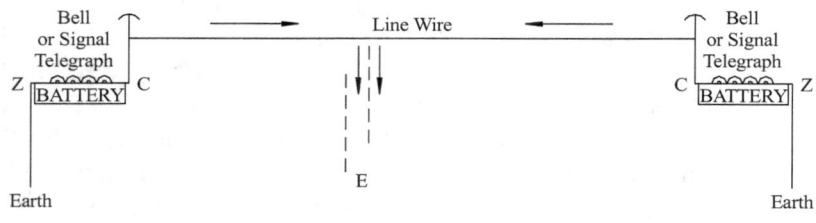

Fig. 103
Train signalling arrangement showing current equilibrium

diagram, Figure 103, shows the effect of a leakage of electricity from any cause at **E**, allowing a flow of the current to take place from the batteries at both ends. There is thus a double advantage, as the apparatus acts as a tell-tale upon itself, and also enables a guard or passenger, by merely cutting the signal-wire at any point between the telegraph stations, to announce at once, in both directions, any casualty; as the train-signalling apparatus at each end of the section would be instantly thrown out of order till the wire was joined together again, after assistance arrived or the train was enabled to proceed.

The receiving (or indicating) part of the instrument should be so arranged that the warning given shall remain as a permanent record till removed by the station giving the warning.

A distinct and simple form of signal is another requirement, and during the last few years a great improvement has been effected in this direction by the actual incorporation of the ordinary railway semaphore signals into the electric apparatus; so that those in charge of the out-door signals and points are warned by the same signals in the telegraph instrument — signals with which every railway station-master, pointsman, or porter is necessarily conversant. The advantage of this method is apparent, even apart from the trouble saved in teaching the railway employees any novel telegraph signals.

The other conditions to be embodied, as far as practicable, are, economy in number of wires, low battery power, easy regulation, and strength of apparatus, so that it may bear working by the heavy hands of railway pointsmen. In some apparatus the signals are supplemented by bells either in connection with the same, or a separate wire.

171. The first train-signalling instrument I shall describe is that of Mr. C. V. Walker, Telegraphic Engineer to the South-Eastern Railway.

The exterior or indicating part of the apparatus is shown in Figure 104, and presents to the eye of the railway employee a railway semaphore in miniature. The shaded or warning arm is regulated by a very simple arrangement of electro-magnets

Fig. 104
Walker's electro-magnetic telegraph semaphore

within, and is only moved and fixed by electric signals from the other station upon the same section. The position of the left-hand arm, shaded (red), is made to represent from time to time the state of the line of rails to which it belongs — whether "clear" or "blocked" — in precisely the same language as the out-door signal; so that a signal-man who understands the one understands the other. The other (white) arm moves with the signalling key of its own instruments, when sending warning to the other station as to the state of the down line. The semaphore arms have but two motions: when up, to indicate a train on the line; when down, that the rail is clear. No signal-man can alter the position of the red arm on his own instrument, and it is put up behind a train by the next station, and lowered as soon as the train has passed that station. Until put down by the other station, no other train on the same line of rails is allowed to pass on to the section. The signal-currents are set in motion by a simple form of contact-key.

The internal arrangement of the indicating apparatus is

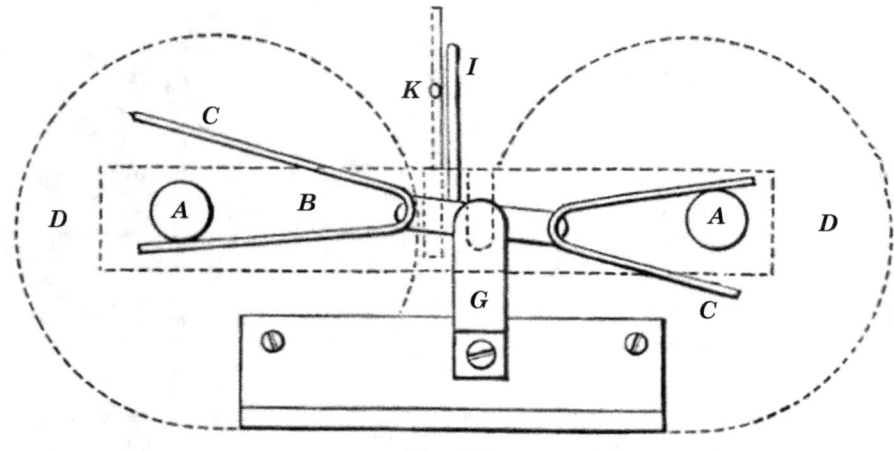

Fig. 105
Walker's semaphore railway telegraph indicating apparatus

shown in Figure 105

Two v-shaped permanent magnets, **C C**, are pivoted on the bearing **G**. A to-and-fro motion is given to them by the poles, **A A**, of the electro-magnetic coils, **D D**. The pin **I**, which is moved by the magnets, acts upon the semaphore arms, and is locked by the stud **K** on the independent armature **B**. The arrangements are so made that the arms are firmly locked in the position in which they have been placed by the electric current, that they cannot be displaced by vibration from passing trains, or external violence, whether accidental or not. The receipt of the reverse current from the other station can alone move them.

In connection with the semaphore there are signal-bells, by which attention is first called, and the state of line or nature of train indicated by a simple system of beats, before the semaphore signal is changed.

This form of train-signal instrument combines the advantages of simplicity and economy, only one wire being required to signal the state both of the up and down lines on any section of railway. The method has been put to the severest tests on the South-Eastern and other railways, particularly where the traffic is heaviest and most intricate, as in the approaches to the London

Bridge Terminus and the Charing Cross and Cannon Street Stations.

I will now quote a few important rules laid down by the South-Eastern Railway to preclude two trains running on any signalling section of the rails in the same direction at one time:–

"Every train or engine must be signalled out to the next station before it leaves or passes a station." So that when the business of the day, for instance, commences, Station B. knows that train No. 1 is asking permission to come to him from Station A.; and, to prevent all misunderstanding — "The train or engine must not be started, or allowed to pass, until the next station has *taken* the *out* signal." It is not enough for the first station to *give* the signal — the other station must *take* it; for, "no signal given by one station is complete until taken by the other station repeating it;" by which process a clear understanding is established between the two signallers, that the precise signal sent by one is received by the other. The next rule applicable is, that "every train or engine that *arrives at* or *passes* a station is to be immediately signalled back *in* to the last station;" and it follows, from what has been already stated, that "no second train or engine is to be allowed to follow till the *in* or *arrival* signal of the *previous* train has been taken." †

It is also provided that the *in* signal is only to be given when the tail-lights have actually been seen, or the guard of the train spoken to, so as to avoid the chance of any part of the train having become detached and left behind on the section over which the train has just been signalled.

172. A somewhat similar mode of signalling was devised by Mr. Preece in connection with the South-Western Railway. A sketch and section of his arrangement is shown in Figures 106 and 107.

The principal difference between this plan and Messrs. Walkers' is found in the electrical portion of the apparatus. To signal both up and down trains, Mr. Preece employs three wires and separate instruments, while Messrs. Walker need but one. He claims this as an advantage, in point of safety, upon the following grounds:–

"When three wires are employed, the all clear signal is main-

† "Train Signalling" by C. V. Walker, F.R.S.

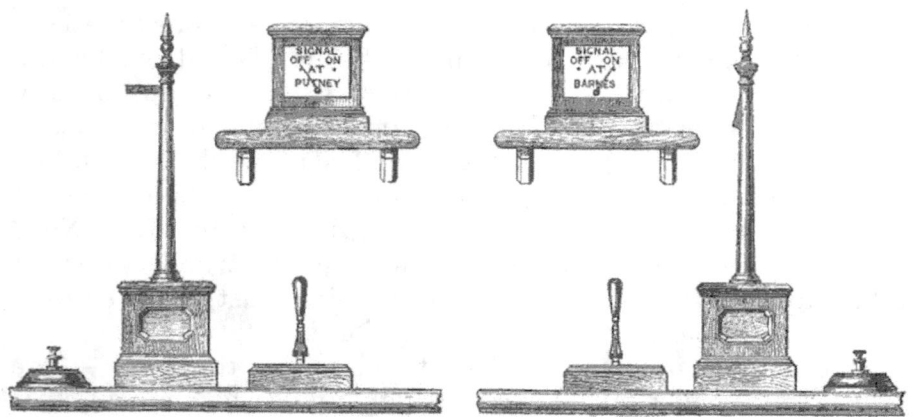

Fig. 106
Preece's semaphore telegraph

tained by a constant and permanent flow of electricity. The result is, that if this current be interrupted or stopped by any cause, the arm will at once fly up, and the danger signal be exhibited. When one wire only is employed, the necessary signals are produced by momentary currents of electricity, and the permanence of the signal is maintained by local magnetism or gravity — the result is, that no derangement of the apparatus or wire is reported, and the instrument may continue to show safety when danger really exists.

"A signal is shown upon an instrument worked on one wire, and actuated by momentary currents, which is supposed to be an indication of the signal represented at the other station; but this is thoroughly illusory and dangerously deceptive. Such a signal is practically made by the outgoing current before it arrives at the distant station, and it may act correctly upon the one instrument without having affected the other.

"The use of momentary currents necessarily leaves what is termed an open circuit, and renders the instruments peculiarly liable to all the disturbances of atmospheric electricity. A flash of lightning might directly enter the wire, and derange both instruments, but when three wires and permanent currents are employed, the lightning would at once be conducted quietly to the earth, without passing through the semaphore."

There are two batteries at the signalling station — one which sends currents in one direction, and the other which sends the reverse. The working lever of the semaphore is placed in electrical connection by a wire, with the stud of the piston-key. The upper stud *E* of the adjusting points is placed in connection with one battery, the lower stud *F* with the other battery. Now when the working lever rests against the stud *E*, it enables the key to send currents in one direction, which deflect the indicator over to "on" ; and when the lever presses against *F*, it causes the key to send currents from the opposite direction, which cause the indicator to point to "off." It therefore follows, that in whatever position the arm of the semaphore at the receiving station may be, the indicator at the sending station will always correspond with it, with this important reservation that the signal must have been first acknowledged upon the bell; and as it is ruled that every signal sent must be acknowledged, it results that the indicator at the sending station must be a faithful repeater of the signal at the receiving station.

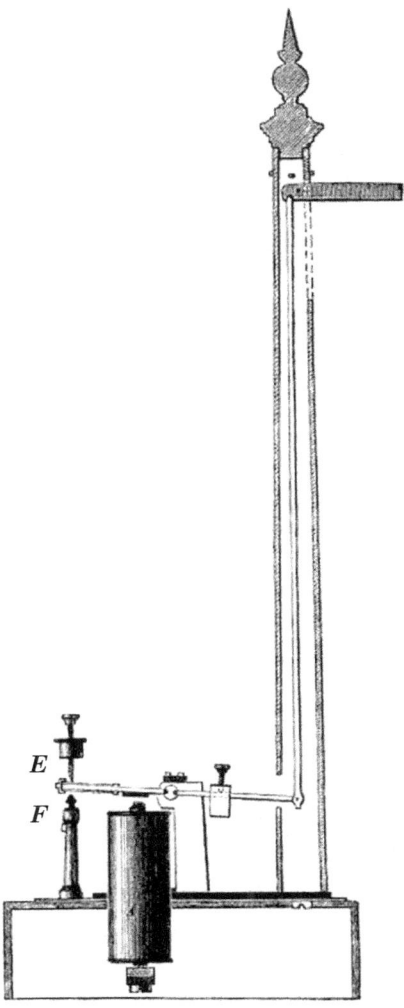

Fig. 107
Preece's semaphore telegraph
(Section)

The necessity for three wires and separate instruments with this apparatus is, however, a considerable drawback in point of economy.

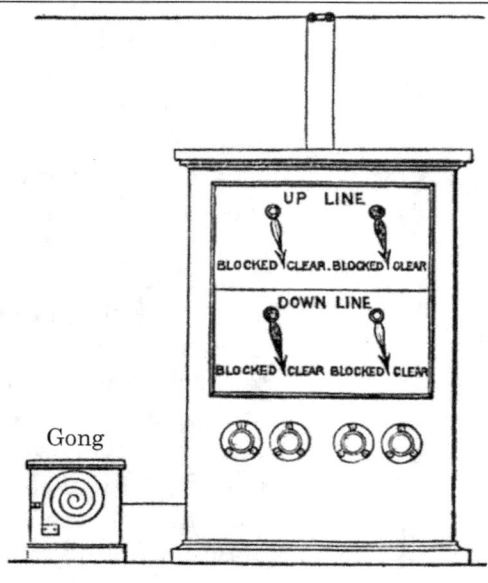

Fig. 108
Tyer's railway telegraph

173. An arrangement for signalling by pointers, as shown in Figure 108, was introduced by Mr. Tyer before the semaphore system was suggested, and is used on several railways.

174. Messrs. Bartholomew and Spagnoletti have each devised train signalling instruments of a similar character as regards the method of indicating, which is carried on by discs, which are made to fill an opening in the dial face of the instrument, as shown in Figure 109.

The instrument is worked by two keys, one for each disc, and labelled the same as the discs, so that by pressing down the white key labelled "line clear," the white disc (line clear) appears in the aperture of the green dial, and on pressing down the red key, labelled "train on line," the red disc (train on line) appears on the green dial; thus one signal only, and the one intended to be shown, can be seen; hence no confusion or doubts can arise. The dial being green, and the discs red and white, by the contrast of colour a clear and unmistakable signal is obtained.

Either of the keys can be pegged down by a pin, which passes

Instrument showing
line clear

Instrument showing
train on line

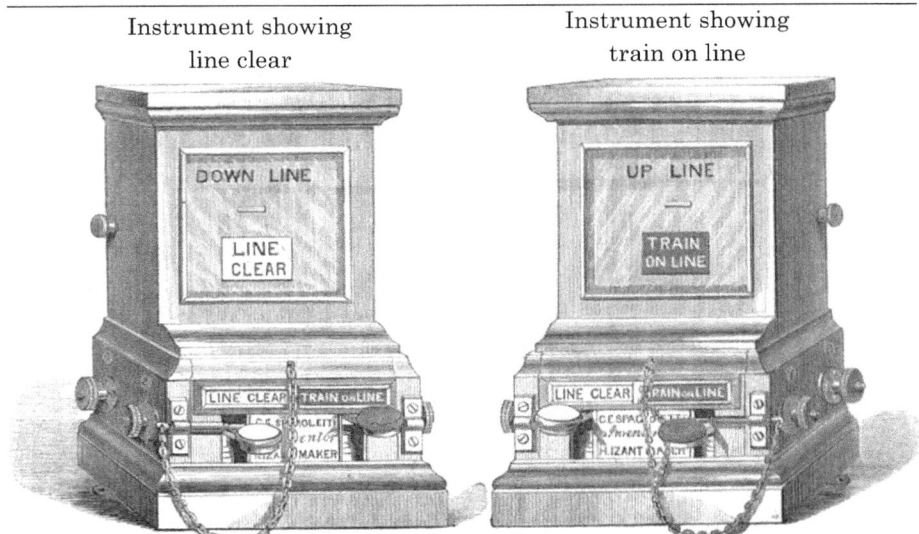

Fig. 109

Spagnoletti's disc telegraph instrument, for signalling trains through tunnels, over single lines, steep gradients, curves, and crowded lines.

through an eye by the side of each key, so that either signal can be kept on as long as necessary. These instruments are so constructed, that whichever signal is given at the station, the same must show at the one to which it is sent; and when a signal is given, and the key pegged down, it cannot be reversed from the end to which it is sent.

This instrument also registers the gradual exhaustion of the battery, so that the battery can be attended to before it becomes too weak to perform its work.

A single striking bell is worked in connection with the apparatus, separately, or attached to the instrument; if the former, it has a key of its own; if the latter, on either of the sending keys being pressed down, the signal is shown, with a simultaneous stroke of the bell; so that by having a code of the number of the beats on the bell, a corroboration is given to the signals shown by the discs.

In this plan three wires are required to work an up and down line; while it does not appear to possess the same advantages as the semaphore form of signal.

175. Various ingenious plans have been introduced recently to provide the means of communication between guards and passengers. Those of Mr. Walker and Mr. Preece are, perhaps, the best known. The principle adopted in each is to ring an electric bell, or give a signal in the guard's van by moving a contact-handle (or bell-pull, as it might be termed) in any compartment of the carriages. A small indicator or signal is shown outside the carriage in Mr. Walker's method, directing the attention of the guard at once to the compartment from which the alarm emanates.

In Mr. Preece's arrangement a thin glass face covers the bell handle in the carriage, and must be broken before the bell can be sounded.

An instance was recently given, showing the use of this precaution to the Exeter express, on the London and South-Western Railway. When approaching Crediton, on the North Devon line, an alarm was raised, and the attention of the guard was immediately attracted to the train; but, observing nothing wrong, the train was not stopped until it reached the next fixed signals. It was there discovered that a passenger had needlessly, if not wilfully, rung the bells. The glass face, which usually protects the handle or bell-pull from the mischievous and curious, was broken, and the handle turned to the position that not only rings the bells, but which so securely locks and fixes it, that the passenger cannot restore it; and therefore, with the broken glass, leads to his certain detection. The passenger's name was taken for the purpose of prosecution, and the train proceeded.

PLATE XVI

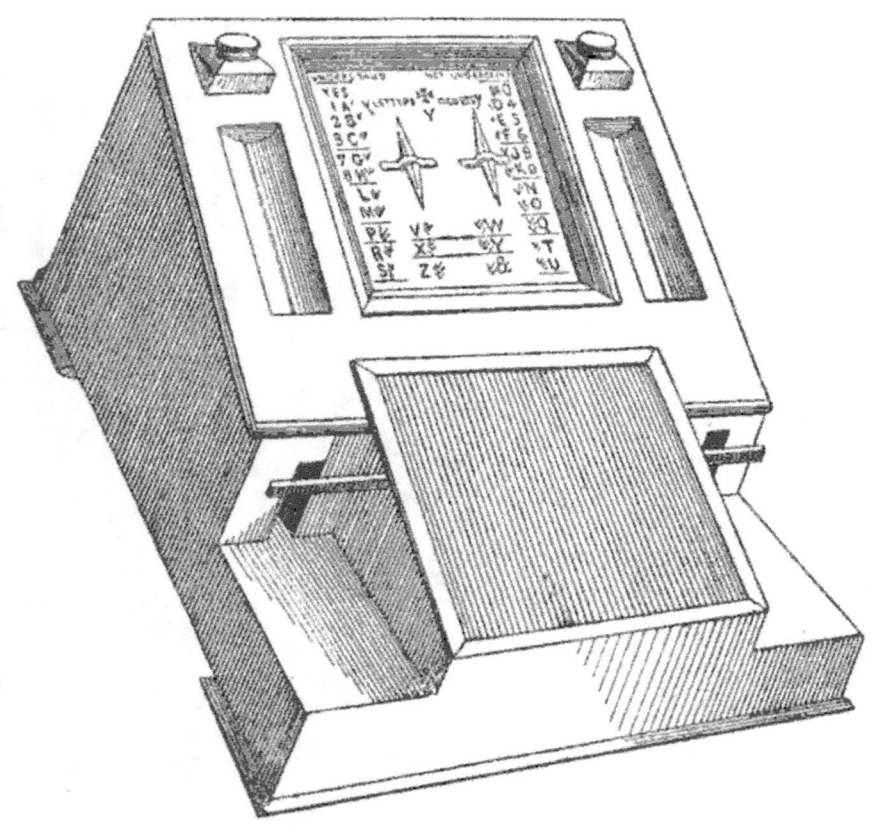

Henley's magnetic needle telegraph

16
GENERAL REMARKS UPON CELERITY OF TRANSMISSION, ETC.

176. Different speed of various telegraphs and operators—177. Conditions upon which celerity depends—178. Speed dependent upon power of receiving apparatus or operator—179. Speed of the Double-needle telegraph—180. Speed of Bright's acoustic telegraph—181. Speed of the Magnetic needle instruments—182. Speed of Morse's and Bain's Telegraphs—183. Transmission of Music—184. Distance sometimes affects celerity—185. Uses of the electric telegraph—186. Subject-matter of despatches—187. Utility of telegraph to railways.

Celerity of Transmission

176. ALTHOUGH it be true that the signals made at any one telegraphic station are rendered instantaneously apparent at another, no matter how distant, it must not therefore be inferred that the transmission of messages by the telegraph is equally instantaneous. Not only is this not the case, but the celerity with which messages are conveyed between station and station, so as to be rendered practically available for the purposes of intercommunication, differs very much when one form of telegraphic instrument or one pair of operators is compared with another.

The profitable result of the operation of any telegraph is evidently measured by the number of words which it is capable of transmitting within a given time, in such a shape as to be intelligible by the person to whom the message is addressed. This, which we shall call the celerity of transmission, and which is quite distinct from the velocity with which electric signals are conveyed from station to station, is therefore a most important element in the estimation of the value of any telegraphic apparatus.

177. This celerity of transmission depends upon a great number of circumstances, several of which are independent of the telegraphic apparatus. The principal of these are:–

1. The skill and agility of the transmitting agent.
2. The quickness of eye, activity and attention of the receiving agent.
3. The instrument used for transmission.
4. The instrument used for reception.
5. The distance to which the dispatch is transmitted.
6. The insulation more or less perfect of the line wires.
7. The weather.

With all and each of these conditions and qualities the celerity with which the dispatches are received and rendered available at their place of destination, varies, and with some of them this variation extends to very wide limits.

178. Different telegraphists have very different powers as to celerity. These powers depend on practice as well as upon natural ability and aptitude, and on manual dexterity. Not only is it necessary to transmit the signals in quick succession, but to do so with such distinctness that they shall be readily interpreted, and such correctness as to render repetitions unnecessary. In this respect telegraphists having equal practice differ one from another as much as do clerks, some writing rapidly and legibly, some rapidly but not legibly, some legibly but not rapidly, and some neither rapidly nor legibly. The relative ability of telegraphists in this respect is partly mental and partly mechanical, depending as much upon quickness of intelligence, attention, and observation, as upon manual dexterity and address.

The great liability to delay, arising from the failure of the transmitter to render himself understood by the receiver, is rendered manifest by the fact that in all telegraphs conventional signs are established for the words, "wait," "repeat," "not understood," "understood," "proceed," and the like. When the transmitter is going on faster than the receiver can take down the words or understand them, then the latter remits the sign to "wait," and if this sign is several times repeated, the necessity of proceeding slower is apparent. If the receiver mistakes a sentence, word, or letter, he remits the sign to "repeat." At the end of each message, he remits the sign "understood," and so on. Now it will be easily conceived that this necessity for frequent interchange of signs between the receiver and transmitter must affect, in an important degree, the celerity of transmission; and that its frequency must depend, not only on the abilities of the telegraphic agents, but also on the character of the signs transmitted by the instruments, according as they are more or less obvious and unequivocal.

It is a remarkable and very curious circumstance, that independently of the mere celerity, clearness, and correctness of transmission with certain telegraphic instruments, each telegraphist has a manner and character, which is so peculiar to himself, that persons receiving his dispatch at a distant station, recognise his personality with as much certainty and facility as they would recognise the handwriting of a correspondent, or the voice and utterance of a friend or acquaintance, whom they might hear speak in an adjacent room. The agents habitually engaged at each of the telegraphic stations, in this way, soon become acquainted with those of the other stations on the same line, so that, at the commencement of a dispatch, they immediately know who is transmitting it.

While the aptitude of the transmitter is partly manual or mechanical, that of the receiver of a dispatch is not at all so. In some telegraphic instruments, as we have seen, the presence of a receiving agent is unnecessary, the dispatch being written or printed by the apparatus itself. In all instruments, however, which merely exhibit arbitrary signals, expressing letters, numbers, or words, the celerity must depend on the skill, aptitude, and quickness of eye of the receiver, to catch and commit to paper the succession of letters or words, as fast as the signals expressing them are produced before him.

In general, it is much more easy to transmit rapidly than to receive rapidly. The transmitter knows beforehand what signs he is about to produce, while each of them comes upon the receiver altogether unforeseen, and if, in the celerity of their succession, one or more of them escape his eye, he is obliged either to guess at the missed letter or letters, which he can sometimes do with all the requisite clearness and certainty, or he must arrest the transmitter, which he does by giving the sign, "repeat," and so delay arises.

In telegraphs which work by a series of visible signs or

sounds, whether they be deflections of the needle, as in the English instruments, beats of muffled bells in the acoustic apparatus, attitudes of the arms, as in the French State instruments, or pointers directed to the letters or figures on a dial, as in the railway instruments, the celerity of the transmission must be determined by the power of the less able of the two agents, the transmitter and receiver. If the transmitter be able to send the letters more rapidly than the receiver can read and take them down, he must moderate his pace to the limit determined by the power of his correspondent. If the receiver be capable of reading and taking down faster than the transmitter is able to send the letters, his superior force is useless. He can only write the dispatch as fast as he receives it. To send dispatches with the greatest advantage of celerity, the agents yoked to corresponding instruments ought to be selected of as nearly equal ability as possible, since the slower of a pair necessarily neutralises the superior skill of his fellow, and the dispatch would proceed with equal celerity if he were yoked with a less able correspondent.

As quickness of hand is essential to the transmitter, quickness of eye is necessary to the receiver.

179. In all forms of telegraph which express the letters by signals, such as the needle acoustic instruments, and the French State telegraph, a certain pause is necessary between letter and letter, to prevent the signals being confounded one with another. In the single needle instrument, the letters being expressed by from one to four deflections of the needle, and in the double needle, from one to two, the mean time of each letter is that of two and a half deflections in the one, and one and a half in the other, the intervals between letter and letter being the same in both. It must, however, be remembered, in comparing the relative celerity of different instruments, that the double needle instrument, as well as the French State telegraph, is, in fact, two independent telegraphs, having not only separate and independent transmitting and indicating apparatus, with their respective accessories, batteries, &c., but separate and independent conducting wires. It is, in effect, as if two equally powerful and independent steam engines were united in the same work, in order to obtain double power.

As long ago as 1850, Mr. Walker made some calculations, with the view to determine the average celerity of transmission at that time with the double needle instrument in the hands of competent operators, and published the results in his work on electric telegraph manipulation. Eleven messages were timed, all of more than the usual length, the shortest consisting of 73 and the longest of 364 words. The total number of words was 2638, and, consequently, their average length was 240 words. The total time of transmission was 162 minutes, and, consequently, the average number of words transmitted, per minute, was 16¼. The greatest speed of transmission was 20½, and the least 8¼ words per minute.

The Secretary of the Electric Telegraph Company, Mr. Foudrinier, also caused a sufficient number of messages, transmitted in the ordinary course of business with the double needle instrument, to be timed, and the following were the results:–

11 Messages. — Number of words in the addresses 84
     "   "   "   messages ....<u>160</u>

              Total number of words transmitted ....<u>244</u>

Total time of transmission ....689 seconds.

Average number of words transmitted per minute 21¼

It appears, therefore, that the average celerity of transmission with this instrument has increased in the ratio of about 16 to 21.

The greatest celerity of transmission was, in this case, 24½, and the least 16¾ words per minute.

180. In the acoustic telegraph, where the signals are produced by short quick blows upon two muffled bells of a different pitch, the speed of transmission is found in practice to exceed either the double needle instrument, or the Morse and other recording telegraphs. In the needle instruments the signal is given to the eye, which is a much less flexible organ than the ear; and in the recording apparatus the signals are much longer in duration on an average, the distinctions being made by differences of duration. The average speed of the acoustic telegraph is between thirty and forty words per minute.

181. The manner in which the magnetic electric current affects the needle in the arrangement adopted by the Magnetic Telegraph Company, being somewhat different from that produced in the common needle instruments, worked by the Electric Telegraph Company, although the systems of telegraphic signals are not essentially different, it appeared probable that the difference between the instruments might more or less affect the celerity of transmission. Mr. Bright, the Secretary of the Magnetic Company, some years ago timed a series of dispatches transmitted in the ordinary course of business, and the return showed:–

74 Messages. Total number of words 2792
Time of transmission 102m 6s
Average number of words per minute.............................27 $\frac{1}{3}$

The greatest celerity of transmission attained in this series of messages was 37 $\frac{1}{6}$ words per minute.

The entire series consisted of messages transmitted from London to Liverpool, on a pair of double needle instruments, at different times of the day, and were carefully tabulated. In the series, several messages were included, the transmission of which was exceptionally slow, owing either to the difficult nature of the communications, consisting of long words in private cipher, or of the names of foreign towns, or, in fine, from the inaccuracy or slowness of the transmitting clerk in London. It would seem, therefore, that this series of messages includes fair conditions for an average result.

It would, therefore, appear that the needle instruments worked by the magneto-electric current used by this company are, *cœteris paribus*, susceptible of greater celerity of transmission than the instruments in which the needles are affected by the common voltaic current, in the ratio of about 27 to 21, or 9 to 7.

One of the causes which has been assigned to this increased efficiency, is the fact that the needles of the magnetic instruments have a dead beat, while those of the voltaic instruments, in striking the stops, have a recoil, and vibrate two or three times before they come to rest. Whether this be the real cause of the difference, further experience must prove, but it is difficult to imagine that it can be due to any cause independent of the instruments, seeing the large number of messages from which the average has been computed.

182. The writing and printing telegraphs are independent of a receiving agent, the receiving apparatus in all these being automatic. Those who print the dispatch in the common letter press characters have the further advantage of being wholly independent of the skill of any agent to interpret or decipher them.

The celerity of transmission attainable with the Morse telegraph, which of all the forms of telegraphic apparatus hitherto invented is the most extensively used, is very considerable, but varies perhaps still more than the needle instruments, with the skill of the telegraphist.

In this instrument, it will be remembered that the transmitting agent plays upon a key-commutator, the letters being severally expressed by repeated touches of the key succeeding each other, after longer or shorter intervals. At the station receiving the dispatch, the armature of the electro-magnet moves simultaneously with the transmitting key, and at each of its motions towards the magnet it produces a distinctly audible click. The receiving agent acquires by practice such expertness and quickness of ear, that by listening to this clicking he is able to interpret the dispatch, and to write it down or dictate it to a clerk without using the apparatus for impressing it upon the paper ribbon.

Different telegraphists acquire this power of oral interpretation of the dispatches with different degrees of facility and precision; but all are more or less masters of it. So much so, that in most cases on the American lines, it is by the clicking of the magnet that the messages are taken down, being afterwards corrected, if necessary, by comparison with the indented paper ribbon.

The telegraphist is placed at a table, upon which the instrument stands, having before him the paper upon which the message is to be written, and at his left a provision of black-lead pencils ready cut and pointed, usually half a dozen. When the transmission of the message commences, the electro-magnet dictates it to him, letter by letter, at the same time indenting it upon the paper ribbon. He writes it down, and, in general, it is delivered by the magnet as fast as he can write it, availing himself of all such abbreviations as are intelligible to those who may have to read it.

As the points of the pencils are successively worn he lays them on the table at his right hand. A person engaged exclusively in that process, visits his table from time to time, re-points the pencils lying on his right, and replaces them on his left. This person passes round the telegraph office, from table to table, keeping up a constant supply of properly pointed pencils at the hand of each telegraphist.

The most expert telegraphists are able to take down the messages in this manner by ear, without any reference to the ribbon, and so correctly that there is no need of subsequent verification. When the message is concluded, the sheet on which it is written is handed to another clerk, who is provided with a stock of envelopes, in one of which he encloses it; and, writing the address upon it, delivers it to a messenger, who forwards it to the party to whom it is addressed. Meanwhile the paper ribbon, on which the message has been indented in the telegraph ciphers, is cut off, folded up, and preserved for reference.

It is only, however, the most expert class of telegraphists that can operate in this way. Others, less able, are always obliged to verify and correct what they have taken down, by comparison with the indented ribbon, after the message has been concluded; while others, less able still, cannot trust themselves to take down by ear, and sit before the ribbon as it is discharged from the roller, writing out the message from it by eye.

The salaries allowed to different agents vary according to the skill they attain in these operations. One who acquires the power of taking down rapidly and correctly by ear will receive twice the amount allowed to him who can only take down by eye, the latter being always much slower than the former.

It often happens that the power of interpreting easily and correctly by ear is very important, as in the case in which the mechanism of the instrument for moving and indenting the paper may have been accidentally deranged and disabled, or in which the office may be deficient in its supply of paper ribbon.

By the aural method of reception the entire receiving apparatus, except the electro-magnet and its armature, is dispensed with.

If a mistake is committed by the transmitting agent, in consequence of which a word or phrase is unintelligible, the receiving agent intercepts the current, and signifies that the word is to be repeated, and at the same time tears off the erroneous part of the ribbon. This, however, is a circumstance of rare occurrence.

When a very long dispatch is transmitted, and arrives with greater celerity than that with which an agent can transcribe it, the ribbon may be divided, and two agents put to work at once at its transcription. The reports of Congress and public meetings transmitted to the journals, afford examples of this.

These reports may be, by one operation, transmitted to all the towns upon the same telegraphic line. In some cases long dispatches, such as those addressed to the journals, are expedited by two or more instruments on different wires. The dispatch is, in this case, divided into two or more parts, marked 1, 2, 3, or A, B, C, &c., and these parts are simultaneously transmitted to their destination, being reunited there after their arrival. This expedient, however, can only be resorted to where there are two or more line wires, which is a rare case in the United States.

All that has been said above relating to Morse's telegraph may, *mutatis mutandis*, be applied to other telegraphic instruments, which write in cipher the dispatches by self-acting machinery.

When dispatches are transmitted by means of a key-commutator, with Bain's telegraph, the operation being precisely similar to that of Morse, the celerity of transmission by operators of equal skill ought to be the same. Nevertheless, when the Bain system was used on several lines by the Electric Telegraph Company, many years ago, a series of messages were timed, of which the following is the summary of the results:—

63 Messages. — Total number of words in the address 456
       "   "   "   messages .....<u>991</u>
         Total number of words transmitted.......<u>1447</u>
Total time of transmission................................ 4454 seconds.
Average number of words transmitted per minute 19½

It appears, therefore, that as this telegraph is worked in England, its rate of transmission is slower than the double needle, or acoustic telegraphs.

183. It is an amusing fact, that music has actually been transmitted in this way by means of its rhythm. The following is related by an eye-witness of the experiment at New York:–

> "We were in the Hanover Street office when there was a pause in business operations. Mr. W. Porter, of the office at Boston, asked what tune we would have. We replied "Yankee Doodle," and to our surprise he immediately complied with our request. The instrument commenced drumming the notes of the tune as perfectly and distinctly as a skilful drummer could have made them at the head of a regiment; and many will be astonished to hear that "Yankee Doodle" can travel by lightning. We then asked for "Hail Columbia!" when the notes of that national air were distinctly beat off. We then asked for "Auld lang syne," which was given, and "Old Dan Tucker," when Mr. Porter also sent that tune, and, if possible, in a more perfect manner than the others. So perfectly and distinctly were the sounds of the tunes transmitted, that good instrumental performers could have had no difficulty in keeping time with the instruments at this end of the wires."†

That a pianist in London should execute a fantasia at Paris, Brussels, Berlin, and Vienna, at the same moment and with the same spirit, expression, and precision as if the instruments, at these distant places, were under his fingers, is not only within the limits of practicability, but really presents no other difficulty than may arise from the expense of the performances. From what has been just stated, it is clear that the *time* of music has been already transmitted, and the production of the sounds does not offer any greater difficulty than the printing of the letters of a dispatch.

184. Although the distance to which the dispatch has to be sent cannot be said directly to affect the celerity of transmission, there are circumstances nevertheless which in practice render the transmission to great, slower than to lesser distances. Long direct circuits are, as a rule, only kept up where there are a sufficient number of messages between the two points to keep them well employed throughout the day. In other cases messages are passed

---

† Chambers's Papers for the People, vol. ix. No. 71.

on from one station to another, and have occasionally to wait on their journey till a wire is clear of prior work, and it is their turn to be sent on a further stage.

And even though it may be practicable to establish a direct communication between two very distant stations by putting the wires in immediate connection, more or less delay must necessarily take place. The telegraphist who transmits, must first send a message along the line to all the intermediate stations to require the wires to be united for direct communication. At these intermediate stations, the wires may be employed, and the message must wait until they are free.

Thus, although it be true that so far as the electric fluid and the apparatus by which it is transmitted are concerned, they are capable of sending a message from pole to pole in an inappreciable interval, yet the machinery of telegraphy as practically constructed presents causes of delay which prevent in many cases this vast celerity from being called into action.

Besides these causes of delay, there are, however, others. It has been stated that the intensity of the current is diminished, *cœteris paribus*, as the distance is augmented. When transmission therefore to great distances is required, various expedients, at intermediate stations, such as relay batteries or relay magnets, or both, are required, and notice must be given to apply these even when they are provided.

The chances of interruption by reason of defective insulation or accidents to the wires, are also increased in proportion to the distance. As may be naturally expected, examples of direct telegraphic communication to great distances are supplied by the United States.

On the lines of the O'Reilly Company of New York, messages are daily transmitted without any intermediate repetition to a distance of 1100 miles, that is from New York to Louisville in Kentucky.

"To do this, it is found necessary to place two relay instruments and batteries in the circuit at a distance of 400 miles apart, for the purpose of renewing the electric current, part of which .escapes from defective insulation and atmospheric causes. By similar arrangements of relays with intermediate batteries at several intermediate points, New Orleans and New York are placed in instantaneous communication with each other. To enable this to take place, requires, in the first place, a line substantially built and thoroughly insulated. It may be remarked, that it is but a few years since, when to telegraph 300 miles on a single or unbroken circuit, was considered a feat; now, from improvements made since then in telegraphs, we can send over 1100 miles easier than we could 300 at that time. At first, in many telegraph offices in the United States, they used a separate battery for each line. From a series of experiments made, one single battery, of no greater strength than those formerly used, now works eight distinct and separate lines, with no apparent diminution of strength, and at a great saving of expense to the office."†

The directors of the lines at New York report that their telegraph messages have in some cases been actually transmitted without intermediate repetition to a distance of 1500 miles, and in the same way London is frequently connected direct to Vienna, St. Petersburg, and occasionally even Odessa; while on the Atlantic cable direct signals have been passed through 3700 miles without relay.

The promptitude with which dispatches are expedited, and the celerity with which they are transmitted, will be greatly promoted in all cases by a uniform system and organisation being established upon the lines over which they are transmitted. No greater cause of delay can exist than that which arises from diversity of telegraphic instruments and language. Much inconvenience, expense, and delay also arise even in cases where similar instruments and ciphers are used, from a want of uniformity in the various parts of the apparatus, and in the systems of abridgments which are adopted in the language. Where the instruments and the parts of apparatus have been constructed of varying patterns and sizes, they cannot be readily replaced in cases of wearing out or accidental fracture. By the adoption of one uniform size and pattern, depots of all the parts may be provided, from which any station which may be .stopped by an accident can be immediately sup-

† Report of Mr. O'Reilly. Jones's El. Tel., p. 101.

plied with the part or parts which require to be replaced. Another advantage incidental to such uniformity is greater economy in the maintenance of the apparatus and lines.

The batteries invariably used by the American telegraphs are those of Grove, each element of which consists of a cup of unglazed earthenware, placed in a glass tumbler of equal height and greater diameter. A zinc cylinder is let down between the glass and the earthenware cup, and a platinum cylinder is let down into the earthenware cup. The space between the cups is then filled with acidulated water, and the earthenware cup is filled with pure nitric acid. These batteries are, however, of a much more expensive class than those used in England.

THE USES OF THE ELECTRIC TELEGRAPH.

185. To form an estimate of the uses which the electric telegraph subserves, it would be necessary to obtain a report of the subjects of the messages classified, with the relative number of each class, which are transmitted from and received by the chief telegraphic stations.

The prevailing subjects of the dispatches vary according to the station from or to which they are sent. Thus, as might naturally be expected, in large commercial marts, such as Liverpool and Glasgow, they are chiefly engrossed by messages of mercantile firms and business. Their prevailing subjects also vary much with the season of the year. Thus, in summer, the messages of tradesmen are greatly multiplied in consequence of the number transmitted by dealers in perishable articles, such as fish, fruit, &c., which must be supplied in regulated quantities with the greatest promptitude.

The following classification of 1000 dispatches passed by the British and Irish Magnetic Telegraph Company from Liverpool in February, 1867, has been made. The comparative numbers vary, however, with the time of the year; for instance, the number of betting messages would be largely increased in the Racing Season:–

General merchants ------------------------------------ 42
Stock and share transactions --------------------- 147
Ship insurers, brokers, &c ------------------------ 170
Banking messages -------------------------------------- 9
Corn dealers -- 21
Betting -- 1
Personal and domestic ---------------------------- 124
General brokers -- 131
Tradesmen --- 40
Cotton brokers, &c ---------------------------------- 295
Law --- 14
Political --- 6

$$\overline{1000}$$

186. Mr. Walker has given the following list of the subjects of telegrams, as a specimen of the uses which the public make of this mode of communication.

Accidents	Customs	Markets	Post-horses, &c
Announcements	Deaths	Medical aid	Remittances
Appointments	Departures	Meteorology	Reporters
Arrests	Dispatches	Missing trains	Respite
Arrivals	Elections	Murders	Robberies
Bankers	Elopements	News	Royal movements
Beds	Expresses	Nurses	Sentences
Bills	Funds and Shares	Orders	Shipping news
Births	Government	Passengers	Ship-stores
Commotions	Health	Payments	Turf
Corps	Hotels	Police	Witnesses
Counsel	Judgments	Political	Wrecks
Couriers	Lost luggage		

Cipher or code has been found of great use in preventing errors, especially where messages have to pass through the hands of foreign clerks. Words of a very simple and easy class may be used and yet each so unlike in spelling as not to be readily confounded, thus obviating in a great measure the mistakes that frequently arise in the transmission of figures, fractions, &c. The code system allows of great condensation of messages, single words being made to express whole sentences. In composing code messages for transmission through Foreign countries it is well to use names of towns, &c., well known abroad. The British Government dispatches are usually formed by groups of figures.

The obstacle to the extension of the uses of the telegraph, created by the tariff, has been of late greatly lessened by the considerable reduction of the prices of transmission, and it may be hoped that ere long the companies and the public will discover that the interest of the one and the convenience of the other will be best promoted by a still further reduction of price, and a still larger use of this mode of intercommunication.

It is probable and desirable that something approaching to the uniform postage system may eventually be realised in the telegraph. The British and Irish Magnetic Telegraph Company are now trying the effects of a 6d. charge for short distance messages on their line between Dublin and Kingstown, and between Cork and Queenstown, &c; and the London District Telegraph Company charge 4d. per message between the offices they have opened in conjunction with the Magnetic Company in London and the suburbs.

187. In the management of railway business in all countries, but more especially upon our own ever-crowded and over-worked lines, the telegraph has become an indispensable accessory, without which this mode of locomotion would be deprived not only of its efficiency but its safety. Consequently the railways in most countries have been provided with lines of telegraph expressly and exclusively for their own use, independently of those which are appropriated to the public service; and on the continent, as already explained, such telegraphs are usually alphabetic, that is, such as convey their messages by pointers, which are successively directed to the letters of the words, so that any of the railway officials who can read, may be able, though slowly, to interpret a message which arrives, or to transmit one to a distant station.

To illustrate the vast utility of the telegraph to the railway,

Mr. Walker gives the following as a classification of their subjects:—

			Messages
1.	Concerning	Ordinary trains	1468
2.	"	Special trains	429
3.	"	Carriages, trucks, goods, sheets, &c. ...	795
4.	"	Company's servants............................	607
5.	"	Engines..	150
6.	"	Miscellaneous matters	162
7.	"	Messages forwarded to other stations .	499
		Total	4110

On the extended systems of railway telegraphs now constructed, a vast number of messages are transmitted in conducting the traffic. On important and widely-spread lines, like the London and North Western, South Eastern, Lancashire and Yorkshire, &c., &c., as many as 300,000 messages are sent by each in a year. In many cases the guards of railway trains are now provided with portable telegraphs. By these the conductor of a train can, whenever the train is stopped between stations, whether from accident or other cause, give immediate notice to the preceding and succeeding stations; so as to prevent a collision by a following train overtaking that which is accidentally stopped; or if necessary he can call for an engine to carry on the train, or any other aid that may be required.

Notices of the passing, starting, and arrival of trains are however transmitted from station to station, quite independent of any accidents that may arise, so that all the station-masters, so far as relates to the movement upon the line, are endowed with a sort of omnipresence; so conscious are they of the possession of this power and its value, that their language is that of persons who actually see what is going on at vast distances from them. Thus, they are in the common habit of saying — "I just saw the train pass such or such station," fifty miles distant perhaps, when in reality all he saw was the deflection of the needle of their telegraph.

"If trains are late, the cause is known; if they are in distress, help is soon at hand: if they are heavy, and progress but slowly, they ask and have more locomotive power either sent to them or prepared against their arrival; if there is anything unusual on the line they are forewarned of it, and so forearmed; if overdue, the old plan of sending an engine to look after them has become obsolete — a few deflections of the needle obtain all the information that is required." †

The utility of special trains is well known. News of the utmost importance, or a government courier bearing dispatches of the greatest urgency, arrives at one of our ports and demands a train instantly for conveyance to London. Now in such cases it does not often happen that a disposable engine is found at the station where the demand is presented; but the telegraph sends a dispatch along the line, calling one from the nearest station at which one can be found, and when the engine has been obtained the special cannot start with safety unless the line is cleared for it.

The telegraph again interposes its aid, and sends a notice along the line of the moment of starting, from which, combined with the known speed of the train, the exact moment when it will pass every station upon the line is known, and of course the line is cleared for it, and all danger of collision removed. How frequent are the occasions for appealing to the telegraph for this aid without which special trains would not only be less rapid, but infinitely less safe, as well for themselves as others, may be seen by reference to the analysis of dispatches we have given above, from which it appears that in three months, upon the South-Eastern lines, there were not less than 429 messages respecting special trains, that is at the rate of about five per day.

In the general management of the traffic upon an active line of railway, an incalculable amount of capital and current expenditure is saved by the telegraph. Without it rolling stock would require to be provided in much greater quantity, and a far greater unprofitable wear and tear by useless trips, of what in railway language are called "empties," would take place. By the telegraph, as we have stated, each stationmaster is ubiquitous so far as the line is concerned. He knows where carriages, wagons, trucks, sheets, and engines are to be found, and how many of them, and he calls

† "Telegraph Manipulation," p. 84

by the telegraph so many, and no more than he wants, and at the time he wants them, from the nearest or most convenient station where they are to be obtained.

Before the establishment of the telegraph, some of these objects were imperfectly attained by means of pilot engines, that is engines taking no vehicle, which habitually run along the line to carry messages from station to station. As an evidence of the immense saving effected by the telegraph in the practical working of railways, on some railways the cost of maintaining and working a single one of those pilot engines, (all of which have been superseded by the telegraph,) amounted to a greater sum than is now required to defray the expense of the entire staff of telegraph clerks, and the mechanics and labourers employed in cleaning and repairing the instruments and maintaining the integrity of the line wires.

PLATE XVII

Early method of coiling cable in the hold of a vessel.

17
USES OF THE TELEGRAPH

188. AMONG the serious railway accidents which might have been, or actually were prevented by the telegraph, the following have been mentioned:–

In a storm, the wind blew a first-class railway carriage, which stood in an open shed at a second-class station, and putting it in motion upon a very level line, sent it flying with accelerated speed to the terminal station. No telegraph at that time existed to warn either the intermediate or terminal stations of the event and the approaching danger. The vehicle was actually *blown* over twenty-one miles of railway, but the trip it thus took occurring fortunately at an hour of the night when little business was going on, it came to rest without any calamitous result.

On New Year's Day, 1850, a catastrophe, which it is fearful to contemplate, was averted by the aid of the telegraph. A collision had occurred to an empty train at Gravesend; and the driver having leaped from his engine, the latter started alone at full speed to London. Notice was immediately given by telegraph to London and other stations; and while the line was kept clear, an engine and other arrangements were prepared as a buttress to receive the runaway. The superintendent of the railway also started down the line on an engine; and on passing the runaway, he reversed his engine and had it transferred at the next crossing to the up-line, so as to be in the rear of the fugitive; he then started in chase, and on overtaking the other, he ran into it at speed, and the driver of his engine took possession of the fugitive, and all danger was at an end. Twelve stations were passed in safety: it passed Woolwich at fifteen miles an hour: it was within a couple of miles of London before it was arrested. Had its approach been unknown, the mere money value of the damage it would have caused might have equalled the cost of the whole line of telegraphs. They have thus paid, or in a large part paid, for their erection.

As a contrast to this, an engine, some months previously, started from New Cross toward London. The Brighton Company had then no telegraphs; and its approach could not be made known. Providentially, the arrival platform was clear; it ran in, carrying the fixed buffer before it, and knocked down, with frightful violence, the wall of the parcels office.†

189. Among the general uses of the telegraph to the public, many examples of the detection of crime are mentioned. It is generally known that the notorious Tawell, after the commission of the murder, started for London from Slough, by the Great Western Railway. Notice of the crime, and a description of his person, however, flew with the speed of light along the wires and arrived at Paddington so much earlier than the murderer himself, that upon his arrival he was recognised, tracked from place to place, finally apprehended, tried, convicted, and executed.

One night at ten o'clock, the chief cashier of the bank received a notice from Liverpool, by electric telegraph, to stop certain notes. The next morning the descriptions were placed upon a card and given to the proper officer, to watch that no person exchanged them for gold. Within ten minutes they were presented at the counter by an apparent foreigner, who pretended not to speak a word of English. A clerk in the office who spoke German interrogated him, when he declared that he had received them on the Exchange at Antwerp six weeks before. Upon reference to the books, however, it appeared that the notes had only been issued from the bank about fourteen days, and therefore he was at once detected as the

---

† "Telegraph Manipulation."

utterer of a falsehood. The terrible Forrester was sent for, who forthwith locked him up, and the notes were detained. A letter was at once written to Liverpool, and the real owner of the notes came up to town on Monday morning. He stated that he was about to sail for America, and that whilst at a hotel he had exhibited the notes. The person in custody advised him to stow the valuables in his portmanteau, as Liverpool was a very dangerous place for a man to walk about with so much money in his pocket. The owner of the property had no sooner left the house than his adviser broke open the portmanteau and stole the property. The thief was taken to the Mansion-House, and could not make any defence. The sessions were then going on at the Old Bailey. Though no one who attends that court can doubt that impartial justice and leniency are administered to the prisoners, yet there is no one who does not marvel at the truly railway-speed with which the trials are conducted. By a little after ten the next morning — such was the speed — not only was a true bill found, but the trial by petty-jury was concluded, and the thief sentenced to expiate his offence by ten years' exile from his native country.

The following additional illustration is taken from an article on the subject which appeared in the "Quarterly Review," extracted from the telegraph book preserved at the Paddington station:–

"Paddington, 10.20 A.M. — 'Mail train just started. It contains three thieves, named Sparrow, Burrell, and Spurgeon, in the first compartment of the fourth first-class carriage.'

"Slough, 10.48 A.M. — ' Mail train arrived. *The officers have cautioned the three thieves*'

"Paddington, 10.50 A.M. — 'Special train just left. It contained two thieves: one named Oliver Martin, who is dressed in black, *crape on his hat*; the other named Fiddler Dick, in black trousers and light blouse. Both in the third compartment of the first second-class carriage.'

"Slough, 11.16 A.M. — 'Special train arrived. Officers have taken the two thieves into custody, a lady having lost her bag containing a purse with two sovereigns and some silver in it; one of the sovereigns was sworn to by the lady as having been her property. It was found in Fiddler Dick's watch- fob.'

"It appears that, on the arrival of the train, a policeman opened the door of the 'third compartment of the first second-class carriage,' and asked the passengers if they had missed anything? A search in pockets and bags accordingly ensued, until one lady called out that her purse was gone. 'Fiddler Dick, you are wanted,' was the immediate demand of the police-officer beckoning to the culprit, who came out of the carriage thunderstruck at the discovery, and gave himself up, together with the booty, with the air of a completely beaten man. The effect of the capture so cleverly brought about is thus spoken of in the telegraph book:–

"Slough, 11.51 A.M. . — ' Several of the suspected persons who came by the various down-trains are lurking about Slough, uttering bitter invectives against the telegraph. Not one of those cautioned has ventured to proceed to the Montem.'

"Ever after this the light-fingered gentry avoided the railway and the too intelligent companion that ran beside it, and betook themselves again to the road — a retrograde step, to which on all great public occasions they continue to adhere."

190. Personal and domestic messages are most generally confined to cases of urgency, and often of distress, painful or ludicrous, as the case may be. Persons in easy circumstances, often resort to the telegraph to gratify a caprice or to obtain some object of gratification for which they are impatient. The mixture of subjects which the agents in rapid succession read from the needles is most curious. " We have," says Mr. Walker, "ordered a turbot, and also a coffin; a dinner, and a physician; a monthly nurse and a shooting-jacket; a special engine, and a chain cable; an officer's uniform, and some Wenham-lake ice; a clergyman and a counsellor's wig; a royal standard, and a hamper of wine; and so on. Passing over the black leather bag which some one every day appears to leave in some train, passengers have recovered luggage of the most miscellaneous character by means of the telegraph. In the trains have been left a pair of spectacles, and a pig; an umbrella, and *Layard's Nineveh*; a purse, and a barrel of .oysters; a great-coat, and a baby; and boxes and trunks, *et id genus omne*, without number."

191. Independently of the direct use made of the electric telegraph by the general public, for the transmission of private dispatches, the several companies have established, in various principal places, news-rooms, where intelligence is from hour to hour posted, as it arrives from all parts of the world.

The Telegraph Company, soon after its establishment, opened subscription news-rooms in the chief towns of England, especially those of the northern counties, in which intelligence of every description which could interest the general public was posted from hour to hour during the day, immediately on its transmission. from London. These establishments did not, however, receive the necessary public support, and with one or two exceptions, they have been discontinued. There is, however, in the Lothbury establishment, besides the private message department, a general intelligence office, in which the news published in the morning journals is condensed and transmitted to the Exchanges of Liverpool, Bristol, Manchester, Glasgow, and other chief provincial centres of business.

During the day, the London and other news is collected, condensed, and transmitted to the offices of upwards of 400 provincial papers, which thus receive during the night before their publication the most recent intelligence of every sort received by telegraph from all parts of Europe, besides the current news of the United Kingdom to the latest moment. The chief of this department, Mr. C. V. Boys, and numerous assistants, are exclusively engaged in the business of this department, and their office presents all the appearances of the editor's room of a widely circulating journal. At six in the morning the news staff are to be seen deep in the "Times" and other daily papers, just hot from the press, making extracts and condensing into short paragraphs all the most important news, which are immediately transmitted to the country papers to form second editions. Neither does the work cease here, for no sooner is a second edition published in London than its news, if of more than ordinary interest, is transmitted to the provinces. Arrived at the chief places in direct communication with London, swifter than a rocket could fly the distance, like a rocket it bursts, and is again carried by diverging (branch) wires into a dozen neighbouring towns of less magnitude and importance.

Besides this organisation for the general transmission of dispatches from the great metropolis to the provinces, there are some curious special arrangements made for the satisfaction of the wants of particular classes. Thus a wire is exclusively appropriated to communications between the Octagon Hall of the Houses of Parliament and the telegraphic station in St. James's-street, the centre of the West-end clubs. This particular wire should be called the "whipper-in" of the House, for it is nothing more than a call-wire for members. The company employ reporters during the sitting of Parliament to make an abstract from the gallery of the business of the two Houses as it proceeds, and this abstract is forwarded at very short intervals to the office in St. James's-street, where it is set up and printed, additions being made to the sheet issued as the MS. comes in. This flying sheet is sent half-hourly to the following clubs and establishments:– Arthur's; Carlton; Oxford and Cambridge; Brookes's; Conservative; United Service; Athenaeum; Reform; Travellers'; United University; Union; and White's. Hourly to Boodle's, Guards, and Army and Navy Clubs; and half-hourly to the Royal Italian Opera. The shortest possible abstract is of course supplied, just sufficient, in fact, to enable the after-dinner M.P. so to economise his proceedings as to be able to finish his claret and yet be in time for the ministerial statement, or to count in the division. On the opposite page, for instance, is a facsimile of the printed abstract:–

The wire to the Opera is a still more curious example of the social services the new power is destined to perform. An abstract of the proceedings of Parliament (see page 286), similar to the above, but in writing is posted during the performance, in the Lobby; and Young England has only to lounge out between the acts to know if Disraeli or Lord John Russell is up, and whether he may sit out the piece, or must hasten down to Westminster. The Opera-house even communicates with the Strand-office, so that messages may be sent from thence to all parts of the kingdom. The government wires go from Somerset-house to the Admiralty, and thence in one direction to Portsmouth and Plymouth by the South-Western and Great Western Railways; and in the other to the naval establishments at Deptford, Woolwich, Chatham, Sheerness, and to the

TELEGRAPHIC PARLIAMENTARY REPORT

House of Commons, Friday, February 8th, 1867

TIME.			TIME.		
H.	M.		H.	M.	
4	0	House made. Petitions for Private Bills. Public Petitions.			attention to the paragraph in the Queen's speech, and to make a statement thereon. He hoped the House would not press for information be-fore Monday next.
4	30	Unopposed Returns. Notices of Motions. Questions. On the motion for adjournment until Monday,			
5	30				Mr. Hardy moving for Bill to establish Asylums for the Sick, Insane, &c. of the Poor of the Metropolis.
5	0	Mr. C. Forster inquiring in-tentions of Government as to the Law of Forfeiture on con-victions of Felony. Mr. Walpole intimated he would bring in a Bill on the subject. Motion for Adjournment agreed to. In answer to Mr. Gladstone, as to whether the Govern-ment could give the House any information as to the mode of procedure on the question of Reform, The Chancellor of the Exche-quer stated the notice he intended to give, was to call	6	0	Still speaking.
			6	30	Ditto. Mr. Ayrton observations. Lord Enfield ditto. Dr. Brady ditto. Mr. Locke ditto. Agreed to.
			7	0	Mr. Walpole moving for Bill to facilitate inquiries by Royal Commissions into Trades' Unions, &c.
			7	30	Still speaking.
			8	0	Sir G. Grey observations. Mr. Hughes ditto. Mr. Goschen ditto. Mr. Roebuck ditto.

Cinque Ports of Deal and Dover. They are worked by a staff provided by the telegraph companies, and the more important messages are usually sent in cipher, the meaning of which is unknown even to the telegraphic clerks employed in transmitting it. In addition to the wires already spoken of, street branches run from Buckingham Palace and Scotland-yard (the head police-office) to the station at Charing-cross, and thence to the city; whilst the Post-office,

CLUB AND OPERA WIRE

TIME.			TIME.		
H.	M.		H.	M.	
8	30	Mr. Neate ditto.			HOUSE OF LORDS.
		Sir F. Crossley ditto.			The Duke of Somerset de-fended the policy of the late Admiralty.
		Mr. Ayrton ditto.			
9	0	The Attorney-General answer-ing objections.			
		Mr. W. Forster observations.			The Earl of Dudley called at-tention to the threatened Political Procession on Mon-day.
		Mr. Kinnaird ditto.			
		Mr. Whalley observations.			
		Mr. Walpole replied.			The Earl of Derby deeply re-gretted the affair, but it was within the law. The Go-vernment would take mea-sures to preserve the peace.
		Motion agreed to.			
9	30	Mr. R. Gurney moving for Bill to remove some defects in administration of the Criminal Laws.			
		Agreed to.	6	50	LORDS ADJOURNED.
		Remaining business disposed of.			
9	40	HOUSE ADJOURNED until 4 o'clock on Monday.			

Lloyd's, Capel-court, and the Corn Exchange communicate directly with the central offices.

The telegraph companies have made combined arrangements by which the correspondents of the press are allowed to forward messages upon an entirely different basis; the charge for intelligence so transmitted amounting to only one-third of the charge to the public, the matter being more voluminous, and in a great measure passing through the wires at a time when they are not otherwise occupied.

The companies also supply the press and news-rooms in various parts of the United Kingdom with news by contract, at the rate of about one halfpenny per line of ten words; and are enabled to do so by making manifold copies of the information (whatever be its nature) for the use of all the press, &c., in each town or district through which such news passes.

Under such arrangements, intelligence to the amount of four closely printed newspaper columns, or more, daily, is trans mitted between all the stations, conveying information of the various share, corn, cotton, coal, iron, cattle, provision, and produce markets; fairs, shipping arrivals, foreign and domestic information, Gazette news, Parliamentary reports, &c. Each piece of news, whatever its nature, obtained in *one* town being conveyed to all the rest: the arrival of vessels in Liverpool, the result of a market in Glasgow, or of a cattle fair at Ballinasloe, affording intelligence for the whole of the United Kingdom, and *vice versa*.

In order to carry out this system, the companies employ paid agents, news collectors, parliamentary reporters, &c., under the direction of Mr. Boys in London.

The following anecdote may amuse, in connection with news-paper telegraphing:– "Four Scotch newspapers have each a special telegraph wire between London and Scotland — the *Scotsman*, and the *Daily Review*, in Edinburgh, and the *Herald* and the *Mail*, in Glasgow — by which important news items are nightly transmitted to the respective journals. .Whether there is any great necessity for such expensive newspaper enterprise, it is not our province to inquire; it is enough for our purpose to say that 'special wiring' has become a feature of Scottish journalism, if it is not also a millstone round its neck. The other night Mr. Moffat, who transmits news from London to the *Glasgow Daily Mail*, arrived late at the Gresham-street telegraph station, and found to his dismay that the gate was shut and bolted, and the night porter sound asleep. Knocking was of no use, kicking had quite as little effect; the porter slept the sleep of the man with an easy conscience. What was to be done? The doughty Scot knew that high up in the building a telegraph clerk waited anxiously for the last of his 'copy,' and he further knew, to his cost, that there were pains and penalties attached to the non-transmission of the 'latest intelli-gence.' At length a bright idea struck him. He ran to Threadneedle-street, telegraphed to Glasgow that the 'special-wire' porter was asleep, and requested the clerk at the Glasgow end to let the Gresham-street clerk know that such was the case. This involved telegraphing nearly one thousand miles, but, nevertheless, within ten minutes the Gresham-street telegraph clerk received the instructions, came down stairs, awoke the slumbering official, and gave admission to the excited journalist and his batch of 'copy!'"

192. It is a fact well known that the electric telegraph is much more extensively used for all purposes, political, commercial, and domestic, in the United States than in this or any other part of Europe. The rates for messages were at one time lower in the United States than in Europe, but now on an average the American charges are considerably higher, while the number of words of an ordinary average are limited to ten, instead of the twenty allowed in this country. We must therefore refer the more extensive use of the wires among our Trans-Atlantic descendants to the great distances between many important towns in America, and to the comparative slowness and uncertainty of their postal communication. In the United Kingdom a letter can be transmitted by post in a few hours, and the Post Office, by its promptitude and punctuality, supersedes the use of the wires in a vast number of cases.

In addition to this, the population of America are much more migratory and speculative, and therefore far more inducements to use the telegraph exist.

We shall notice the question of the tariffs hereafter.

The classes of messages entitled to precedence are government messages, and messages for the furtherance of justice in detection of criminals, &c.; then death messages, which include cases of sickness when the presence of a party is sent for by the sick and dying. Important press-news comes next; if not of extraordinary interest, it takes its turn with the mercantile messages.

Commercial houses resort largely to the telegraph. For example: a person purchasing goods in a market, gives his reference to the merchant — such reference being perhaps 700 or 800 miles away from him. By the aid of the telegraph, the merchant can learn the standing of his customer, even before the purchase is completed. There are bankers, brokers, &c, that receive and send, on an average, ten or twenty messages per day, throughout the year. The telegraph is sometimes used by commercial men to almost as great an extent as the mail. This can be better illustrated by the number of messages sent and received between cities whose commercial relations are intimate, during the hours from 10 a.m. to 5 p.m. For instance: there are transmitted daily, between the cities of London and Manchester, or New York and Boston, 1000 or more messages; three-fourths, if not four-fifths, of which are transmitted between the hours above named. The number of messages sent by a commercial house is governed by the excitement there is in the market of the particular article they may be dealing in. If there are "ups and downs" in the market, money is lavished upon the telegraph freely.

It often happens that a person desires to "converse" with another 400 or 500 miles off. An hour is appointed to meet in the respective offices, and they converse through the operator. Cases may be mentioned of steamboats being sold over the wires — the one party being in Pittsburgh, the other in Cincinnati. Each party wrote down what they had to say, haggled awhile, and finally concluded the sale. Their correspondence was filed away, like other messages, and kept for reference, if ever called in question. It is often used by parties, when from home, corresponding with their families. Sometimes it is the messenger of woe; and anon, that of pleasure. In the early part of 1852, the Astor House of New York, and the Burnet House of Cincinnati, had a series of telegraphic parties. An account of one of them was published in the "Cincinnati Gazette," the parties conversing being about 750 miles apart.

In the same way the chess clubs of Liverpool, Dublin, Manchester, &c., occasionally arrange to meet in rooms attached to the telegraph offices, and play chess matches through the wires, each move being thus dictated.

The following example of the activity of journalism is given by Mr. Jones, who was himself a telegraphic agent for the newspapers:– "Some time back the *Asia* arrived at Quarantine, near New York, about 8 p.m., and was detained an hour by the health officer. The agent of the New York Associated Press and of the New Orleans Merchants' Exchange, Mr. Jones, to gain but a few minutes, had a boat in readiness when the *Asia* brought to. A small bag, containing the latest news, was handed over the steamer's side, to the small boat. By great exertions she gained New York half-an-hour ahead of the *Asia*. The bag was opened — a copy of her news was handed to us, addressed to the Merchants' Exchange, New Orleans, signed 'Jones' — to work we went. It was being transmitted over the wires amid the thundering of the *Asia's* cannon, as she rounded the point; and a complete synopsis of her commercial and political news was received in Louisville, 1100 miles in the interior, before the ship had actually reached the city." In the same way, the steamers from America have been met, for years prior to the establishment of the Atlantic Telegraph, by special telegraph steamers off the coast of Ireland, at Cape Clear, Crookhaven, and Greencastle; and the news brought on their homeward voyages wired to England and the continent of Europe many hours before their arrival in port.

193. The recent attempts of the Fenians to effect a rising in Chester and Ireland were baffled by timely advice given by the wires to the government, troops, and police.

In his reports to Congress, Prof. Morse has supplied various examples of the use made of the telegraph by all classes of persons. During the Philadelphia riots of 1844, the mayor of that city sent an express by railway to the President of the United States at Washington. On the arrival of the train at Baltimore, the contents of the express transpired, and the telegraph, which was then just put in operation between Baltimore and Washington, not being yet established elsewhere in the States, sent on the substance of the dispatch. The President held a cabinet council while the dispatch itself was coming, and had his answer prepared and delivered to the messenger who brought the dispatch at the moment of his arrival, who returned with it instantly to Philadelphia.

194. Medical consultations occasionally take place by telegraph. A patient in or near a country village desires to consult a leading medical practitioner at four or five hundred miles distant. With the aid of the local apothecary, or without it, he draws up a short statement of his case, sends it along the wires, and in an hour or two receives the advice he seeks, and a prescription. Cases are recorded in which electric marriages have been contracted between parties separated one from another by many degrees of latitude. A correspondent of the author of a paper in Chambers's Collection states, that in the United States "the telegraph is used by all classes, except the very poorest — the same as the mail. A man leaves his family for a week or a month; he telegraphs them of his health and whereabouts from time to time. If returning home, on reaching Albany or Philadelphia, he sends word of the hour that he will arrive. In the towns about New York the most ordinary messages are sent in this way — a joke, an invitation to a party, an inquiry about health, &c. In our business we use it continually. The other day, two different men from Montreal wanted credit, and had no references. We said, 'Very well; look out the goods, and we will see about it.' Meanwhile we asked our friends in Montreal, 'Are Pump and Proser good for one hundred dollars each? ' The answer was immediately returned, and we acted accordingly, probably much to our customers' surprise. The charge was a dollar for each message, distance about 500 miles, but much further by telegraph, as it has to go around to avoid water. If my brother goes to Philadelphia, he telegraphs, 'How is the family? What is doing? ' — I answer, 'All well. Sales so much,' and so on."†

195. It has been contended by some, that one of the most serious drawbacks to the general extension of the use of the electric telegraph is the impracticability of preserving that secrecy which the seal confers on written correspondence, the absence of which would utterly annihilate the utility of the post-office. The imperious necessity of guarding this secrecy inviolate, is apparent in the heavy penalties attached to the rupture of the seal, which can only be effected with impunity by a special authorisation of a secretary of state. The vital importance of preserving secrecy in the telegraph is implicitly acknowledged by the heavy penalties, the smallest of which is dismissal, imposed in all countries on the agents who disclose the contents of private telegraphic correspondence.

Secrecy is of course ensured by the use of any species of cipher, but this supposes that the parties corresponding have previously

---

† "Chambers's Papers for the People," vol. ix., No. 71.

prepared the cipher, and are mutually possessed of its key. Such a condition can only be practically fulfilled by correspondents having habitual need of intercommunication, such as mercantile establishments interchanging news of the markets, stocks, sales, and other commercial details; but for the occasional communications of domestic life it is quite unavailable.

If the same privacy as is afforded by the post-office can be thus secured to telegraphic communications; and if by the multiplication of their wires, and the improved efficiency of their instruments, the companies are enabled to reduce their tariff to a still lower limit, and to base it on some uniform principle similar to the admirable penny-postage system of Mr. Rowland Hill; it is difficult to foresee the extent of the revolution which this noble gift of science to mankind may effect. Great as the benefits have been which the post-office has conferred, they will sink to nothing compared with those of the telegraph. In estimating the importance of the part reserved for this vast agent of civilisation, it must not be forgotten that it is still in its early infancy, and that its most wondrous powers are not yet developed by time and growth.

It is, however, only in rare instances that cipher messages are resorted to. The public confidence has been won by the general secrecy observed by the telegraphic agents, and in general no apprehension of disclosure prevents persons from sending the most private and confidential dispatches in the usual manner. One of the directors, who for four years has had the superintendence of extensive lines, states, that in that interval he never heard of an instance of the contents of a dispatch being divulged.

Another circumstance which experience has made manifest has given security to the public on this point. It appears that the agents who are for many hours labouring at the machine in the transmission of dispatches, word by word, rarely are able to give that kind of attention to the sense and purport of the whole which would be necessary to the clear understanding of it. Their attention is engrossed exclusively in the manipulation necessary to transmit letter after letter, and they have neither time nor attention to spare for the subject of the whole dispatch. The case is very analogous to that of compositors in a printing-office, who, as is well known, go through their work mechanically, without giving the least attention to the subject.

A sort of code, or abbreviations, are much in use, however, by mercantile houses. This is practised more for the sake of economy than secrecy, although the latter purpose is also attained. The firm and its correspondents have a key in which are tabulated a number of single words, each of which expresses a phrase or sentence, such as is of frequent occurrence in such communications. The following example of such a commercial dispatch is given by Mr. Jones. The dispatch to be sent consisted of 68 words, as follows:—

"Flour Market for common and fair brands of western is lower, with moderate demand for home trade and export. Sales, 8000 bbls. Genesee at 5 dols. 12. Wheat, prime in fair demand, market firm, common description dull, with a downward tendency, sales, 4000 bushels at 1 dol. 10. Corn, foreign news unsettled the market; no sales of importance made. The only sale made was 2500 bushels at 67 c."

This dispatch, when converted into code, was expressed in nine words, as follows:—

"Bad came aft keen dark ache lain fault adapt."

Complicated systems of code and cipher were invented for the transmission of parliamentary and law reports, and those of public meetings. When the tolls, however, were reduced by competition, this system was abandoned, and the reports were sent in full, or with such abbreviations only as are obvious.

196. In New York, seven of the principal journals formed an association, many years ago, to telegraph in common, sharing the expense. Each journal was, however, at liberty to order for itself any extra intelligence, giving the others, or any of them, the option of sharing it.

Mr. Jones relates that one of the earliest telegraph feats, after the extension of the telegraph lines west to Cincinnati, was brought about by the agency of the "New York Herald," and before any regular association of the press was formed in New York.

"It became known that Mr. Clay would deliver a speech in Lexington (Ky.), on the Mexican war, which was then exciting much public attention. Mr. Bennett, editor and proprietor of the 'Herald,' desired us to have Mr. Clay's speech reported for the paper. We at once proceeded," says Mr. Jones, "to make arrangements to carry it into effect. We had a regular and efficient reporter already employed in Cincinnati, a Mr. G. Bennett; we also had a Mr. Thompson in Philadelphia in cooperation with us for some papers there, and which agreed, if the speech was first received, to share the expense with the 'Herald', the 'Tribune' in New York, and the 'North American' in Philadelphia, agreed to start for a report of the speech, in opposition. From Lexington to Cincinnati was eighty miles, over which an express had to be run. Horses were placed at every ten miles by the Cincinnati agent. An expert rider was engaged, and a short-hand reporter or two stationed in Lexington. When they had prepared his speech it was then dark. The express-man, on receiving it, proceeded with it for Cincinnati. The night was dark and rainy, yet he accomplished the trip in eight hours, over a rough, hilly, country road. The whole speech was received at the 'Herald' office at an early hour the next morning, although the wires were interrupted for a short time in the night, near Pittsburgh, in consequence of the limb of a tree having fallen across them. An enterprising operator in the Pittsburgh office, finding communication suspended, procured a horse, and rode along the line amidst the darkness and rain, found the place, and the cause of the break, which he repaired; then returned to the office, and finished sending the speech."

The Philadelphia "North American," upon whom the "Tribune" chiefly depended, failed to get its report, and the latter purchased a copy from the "Herald."

The expense of securing the speech by express and telegraph amounted to about 100*l*.

197. The telegraphs have derived a considerable share of their revenue from the press. The whole expense, for telegraph reports of all kinds, have some years cost the New York Associated Press about 40,000*l* per annum. Since the opening of the Atlantic

Telegraph one American paper alone, the "New York Herald," has paid for long dispatches at the rate of a sovereign a word. Single dispatches to this paper, giving the King of Prussia's speech after peace with Austria, and particulars of the fight between Mace and Goss, cost over 800*l.* each. The provincial newspapers in England pay the Telegraph Companies in the aggregate about 20,000*l.* per annum for reports of all items of important news. The companies employ their agents and reporters to collect and condense information at various points:– consisting of markets; shipping arrivals and departures; state of weather; political, foreign, and domestic news; abstracts of the parliamentary debates; results of races, &c. This mass of reports averages about four columns daily, and is supplied to the newsrooms and newspapers throughout the kingdom for a small annual payment from each. In some cases the leading newspapers supplement this supply by hiring the use of special wires, through which their own reporters transmit by means of the company's staff many columns of news. On any occasion of great importance, reports *in extenso* are passed through the wires. On one occasion the President's speech was telegraphed in full from Liverpool in three hours, occupying no less than thirteen columns of close print in the London morning papers.

198. The electric telegraph, an offspring of science, has rendered to its parent great and important services.

From the moment that it was discovered that the pulsations of the electric current could, by means of the conducting wires, be transmitted to any distances, its use in the important problem of the determination of longitudes became conspicuously apparent. By reference to Dr. Lardner's tract on Latitudes and Longitudes,† it will be seen that the difference of the longitudes of two places upon the earth's surface is nothing more nor less than the difference of the hour of the day or night, as shown by two well-regulated clocks at the two places. Thus, if while it is 3 o'clock at one place, it is 4 o'clock at the other, the latter is one hour of longitude east, and the former one hour west of the other; or, if it be preferred to express the longitude in degrees, the one place is 15° east or

---

† "Latitudes and Longitudes" in Lardner's Museum.

west of the other.

Now since the machinery of the electric telegraph supplies the means of making all the timepieces of whatever kind, or wherever placed, which are brought into connection with the same system of wires, move in exact accordance; it is capable of making all the timepieces in the United Kingdom move in exact accordance with the standard chronometer of Greenwich Observatory; or, to take a still larger view of the principle, it is capable of governing the movement of all the timepieces of whatever sort, and wherever situated within the range of the vast network of telegraphic wires, which overspreads the European continent; so as to make them move in accordance with any standard timepiece, which may by common consent be adopted as the common regulator.

If such an uniformity of chronometers were established, the longitudes of all places would be determined by ascertaining by observations on the sun, which are always easy and susceptible of great precision, the local time, that is to say, the time which would be shown by a well-regulated clock on the present system. The difference of the two times, that shown by the common standard regulator and that shown by the local clock, would be the difference of longitude between the place in question and the place where the standard regulator would show local time.

In places at great distances asunder, and in different countries, such horological uniformity would, at first, for civil purposes be attended with some inconvenience, since the hour of noon would vary with the longitude. Thus, at a place 15° east of the standard station, the hour of noon would be one o'clock, and at a place 15° west it would be 11 o'clock. Such an inconvenience would, however, only be felt at the moment of the change of custom. It is obvious that it would be as easy and simple to mark the moment at which the sun passes the meridian by 11 or 1, as by 12.

Incidentally to such an horological uniformity would arise, however, the convenience that the hour of noon at all places would express their longitude with relation to the standard station.

The only obstacle to carrying out such arrangements is that of the expense of providing the wires and electric power. By concert between the Astronomer Royal and the several Electric Telegraph

Companies, the Greenwich local time is announced at certain hours of the day, at conspicuous places in different parts of the country, so that navigators who happen to be in any of our ports may avail themselves of these means of regulating their chronometers. We have already explained the method of giving a signal daily at one in the afternoon, by the fall of a large ball upon the dome of the Royal Observatory at Greenwich. This being generally visible from a considerable extent of the river below London bridge, masters of vessels observing it can regulate their timepieces or note their errors. This system of signals is in progress of extension. By means of a galvanic clock at the Observatory, and the conducting wires which connect that building with the telegraph station at Lothbury, hourly signals giving accurately Greenwich time are transmitted to the offices of the company at Lothbury, and in the Strand near Charing Cross. Similar signals are transmitted several times a day to Tunbridge, Deal, and Dover by the wires of the. South-Eastern Company. Signal balls are let fall over the dome of the Telegraph Office in the Strand, the same instant with the fall of the ball over the Greenwich Observatory. Besides this, time-signals are transmitted on the wires twice a day, at 10 in the forenoon, and 1 in the afternoon, directly from Greenwich to various chief stations upon the system of lines of the Electric Telegraph Company.

From the first instant of the laying of the wires connecting the Greenwich Observatory with the stations of the South-Eastern Railway Company and the Telegraph Companies, it was evident that one of the earliest and most useful applications of them would be the determination of the longitudes of several of the principal observatories in the British Isles and on the Continent. During the year 1853, the earliest opportunities were accordingly taken for determining the longitudes of Cambridge, Edinburgh, and Brussels, which was accomplished with complete success, as far as regards the galvanic communications and the observations of the signals at all the observatories. Subsequently the longitude of Valentia, in Ireland, the most western point of Europe, was accurately determined by joining up the Magnetic Telegraph wires between London and that place, a distance by the line of telegraph of nearly 900 miles. The observatories of Greenwich, Brussels, and Paris are now placed in direct electric connection by the submarine cables between Dover, Calais, and Ostend, to the great advantage and advancement of astronomical science.

In the routine of the business of an observatory, the astronomical clock is an instrument in never-ceasing use. A part of almost every astronomical observation consists in noting with the last degree of precision the moments of time at which certain phenomena take place; and so great is the degree of perfection to which the art of observation has been carried, that well-practised observers are able, by the combination of a quick and observant eye and ear, to bisect a second, and even to approach to a still more minute division of that small interval. In order to enable the reader fully to appreciate the benefit which the telegraph has rendered to astronomy, it will be necessary here briefly to explain the manner in which this kind of observation has hitherto been made.

To determine the moment at which the visual ray proceeding from a celestial object has some definite direction, two things are necessary — 1st, to ascertain the direction of such a ray; and, 2ndly, to observe the time when it has such direction. The telescope, with its accessories, supplies the means of accomplishing the former, and the astronomical clock the latter.

If **T T'** (Figure 110) represent the tube of a telescope, **T** the extremity in which the object-glass is fixed, and **T'** the end where the images of distant objects to which the tube is directed are formed; the visual direction of any object will be that of the line **s c** drawn from the image of such object formed in the *field of view* of the telescope to the centre of **c** of the object-glass, for if this line be continued, it will pass through the object **s**.

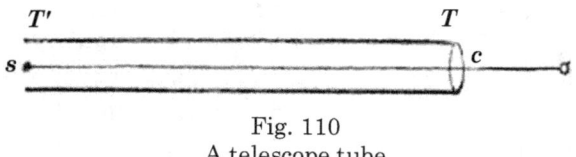

Fig. 110
A telescope tube

But since the field of view of the telescope is a circular space of definite extent, within which many objects in different directions may at the same time be visible; some expedient is necessary by which one or more fixed points in it may be permanently marked, or by which the entire field may be spaced out as a map is by the lines of latitude and longitude.

This is accomplished by a system of fibres or wires, so thin that even when magnified they will appear like hairs. These are extended in a frame fixed within the eye-piece of the telescope, so that they appear when seen through the eye-glass like fine lines drawn across the field of view.

The system consists commonly of five or seven equidistant wires, placed vertically at equal distances, and intersected at their middle points by a horizontal wire, as represented in Figure 111. When the, instrument has been adjusted, the middle wire *m m'* will be in the plane of the meridian, and when an object is seen upon it, such object will be on the celestial meridian, and the wire itself may be regarded as a small arc of the meridian rendered visible. The eye of the observer is occupied in watching the progress of the object moving over the wires in the field of view of the telescope. His ear is occupied in noting and his mind in counting the successive beats of the pendulum, which in all astronomical clocks is so constructed as to produce a sufficiently loud and distinct sound, marking the close of each successive second.

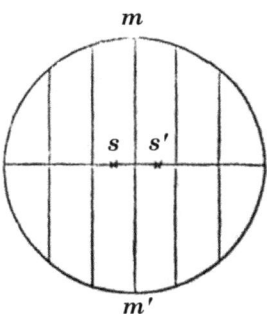

Fig. 111
A telescope wire grid

The practised observer is enabled with considerable precision in this way to subdivide a second, and to determine the moment of the occurrence of a phenomenon within a small fraction of that interval. A star, for example, is seen to the left of the wire *m m'* at *s*, Figure 111, at one beat of the pendulum, and to the right of it at *s'* with the next. The observer estimates with great precision the proportion in which the wire divides the distance between the points *s* and *s'*, and can therefore determine the fraction of a second after being at *s*, at which it was upon the wire *m m'*.

The fixed stars appear in the telescope, no matter how high its magnifying power be, as mere lucid points, having no sensible magnitude. By the diurnal motion of the firmament, the star passes successively over all the wires, a short interval being interposed between its passages. The observer, just before the star approaching the meridian enters the field of view, notes and writes down the *hours* and *minutes* indicated by the clock, and he proceeds to count the *seconds* by his ear. He observes the instant at which the star crosses each of the wires; and taking a mean of all these times, he obtains, with a great degree of precision, the instant at which the star passed the middle wire, which is the time of the transit.

By this expedient the result has the advantage of as many independent observations as there are parallel wires. The errors of observation being distributed, are proportionately diminished.

When the sun, moon, or a planet, or in general any object which has a sensible disk, is observed, the time of the transit is the instant at which the centre of the disk is upon the middle wire. This is obtained by observing the instants at which the western and eastern edges of the disk touch each of the wires. The middle of these intervals are the moments at which the centre of the disk is upon the wires respectively. Taking a mean of the contact of the western edges, the contact of the western edge with the middle wire will be obtained; and, in like manner, a mean of the contacts of the eastern edge will give the contact of that edge with the middle wire, and a mean of these two will give the moment of the transit of the centre of the disk, or a mean of all the contacts of both edges will give the same result.

By day the wires are visible, as fine black lines intersecting and spacing out the field of view. At night they are rendered visible by a lamp, by which the field of view is faintly illuminated.

These points being well understood, no difficulty will be found in understanding the manner in which the telegraph has conferred vastly increased facility and precision on such observations.

The first service which it has rendered is that of making all the clocks in the observatory absolutely synchronous. This has been already accomplished with regard to the solar clocks, that is, those which indicate mean or civil time. It may be, and no doubt will be, also accomplished, with still greater advantage to science, in the case of the astronomical clocks, that is, those which mark sidereal time. The several observers, occupied usually in different rooms, have each their own clock. Now, however perfect may have been the workmanship of these clocks, no two of them can be relied upon to go absolutely together for any length of time; therefore, one of the duties of the observer, and of the conditions of good observations, is to note the error of his clock — that is, its deviation from the standard chronometer of the observatory. These errors are effaced by the expedient of putting all the clocks in the observatory in electrical connection, so that the pendulum of the standard chronometer regulates the pulsations of the current, and these pulsations again regulate the motion of all the other clocks.

The clocks being thus reduced to absolute accordance, the next service rendered by the telegraph to the astronomer consists in affording the means of ascertaining the instant of time at which any celestial object passes across the micrometre wires, with greater facility and precision than were attainable by the use of the eye and ear in the method above described.

This improved method of observation consists in a key commutator placed under the hand of the observer, which governs a current transmitted to an electro-magnet, connected with a style placed over a cylinder coated with paper, upon which it receives a puncture when it is driven down by the pulsation imparted to the current by the finger of the observer acting upon the key. The paper-covered cylinder is kept in uniform revolution at any desired rate by clockwork, and another style, impelled by another current receiving its pulsations from the pendulum of the chronometer, is driven upon the paper with each beat of the pendulum; the interval between two successive marks made by this style representing one sound of time.

Now let us suppose, for example, that by the motion imparted to the cylinder, an inch of the paper passes in each second under the style. The style moved by the clock will therefore leave a succession of marks upon the paper, at distances of an inch asunder. But the particular distance of these marks is unimportant, nor is it material that the cylinder should be moved with mathematical precision. If its motion for the short interval of a second be practically uniform, that will suffice. When the object, a star for example, approaches the field of view, the observer, with his eye to the telescope, holds his finger over the key. He sees the star enter the field and approach the first wire. The moment it crosses the wire, he presses down the key, and the style gives a puncture to the paper on the cylinder. In the same manner, when the star crosses the second and succeeding wires, he again and again presses on the key, and thus leaves as many distinct marks on the paper as there are wires.

After the observation thus made has been concluded, the marks on the paper are examined, and their distances from the preceding and following marks made by the pendulum style are exactly measured, from which is inferred the fractional part of a second, between the moment at which the star crossed each of the wires, and the last beat of the pendulum.

In this way the time of the transit is ascertained to the hundredth part of a second.

The Astronomer Royal, noticing this method of observing in an address delivered before the Royal Astronomical Society, said, that, "In ordinary transit observations, the observer listens to the beat of a clock while he views the heavenly bodies passing across the wires of the telescope: and he combines the two senses of hearing and sight (usually by noticing the place of the body at each beat of the clock) in such a manner as to be enabled to compute mentally the fraction of the second when the object passes each wire, and he then writes down the time in an observing-book. In these new methods he has no clock near him, or at least none to which he listens: he observes with his eye the appulse of the object to the wire, and at that instant he touches an index, or key, with his finger; and this touch makes, by means of a galvanic current, an impression upon some recording apparatus (perhaps at a great distance), by which the fact and the time of the observation are registered. He writes nothing, except perhaps the name of the object observed."

He further observed that it was expected that by this method the irregularities of observation would be greatly diminished; whether because the sympathy between the eye and the finger is more lively than between the eye and the ear, remains to be determined.

It is worthy of remark also, that punctures can be made upon the same revolving barrel by observers employed at two or more instruments erected in different rooms, by means of keys or commutators, which complete the circuit from the same battery to the same puncturing-point. All necessity for comparing clocks is, of course, avoided.

Some difficulties occurred at first in imparting to the cylinder a sufficiently smooth and equable motion, the motion given by common clockwork being always one made by starts like that of the seconds' hand of a pendulum. It was to surmount this difficulty that the Astronomer Royal proposed the substitution of the centrifugal pendulum (resembling the governor of a steam-engine) for the ordinary oscillating pendulum. In the report of the Astronomical Society, it was announced that, "The various difficulties which occurred from time to time in the mechanism of the barrel or smooth-motion clock, used for giving motion to the cylinder on which will ultimately be recorded the transits made with the transit-circle and altazimuth, according to the American method of self-registration, have been overcome. It now carries the cylinders put in connection with it with perfect regularity, its rate having all desirable steadiness."

PLATE XVIII

The Italian terminus of the Mediterranean
submarine telegraph line

18
TELEGRAPH STATISTICS

199. THE rates charged for transmission of messages in the United Kingdom were at first fixed upon a very high scale; but upon the advantage of this means of communication becoming more and more recognised by the public, the tariff was lowered from time to time.

The Electric Telegraph Company, whose lines were first opened in 1847, at the outset derived the greater part of their profits from contracts with railways for the construction and maintenance of railway telegraphs, under Messrs. Cooke and Wheatstone's patents.

The Magnetic and British Companies commenced four years afterwards, and amalgamated subsequently as the British and Irish Magnetic Telegraph Company. Their land lines were linked up to the cables of the Submarine Telegraph Company, which was the first to effect submarine communication with Europe.

While the Magnetic and Submarine Companies have thus extended to France, Belgium, Hanover, and Denmark; the Electric Company, on the other hand, have access to the Continent by means of their cables to Holland.

In 1861, the United Kingdom Company also erected a series of lines between some of the principal towns of the kingdom, and a shilling rate was tried by the three companies for nearly four years; but being found unremunerative, a distance tariff was again resorted to in July, 1865, which still remains in force. The following is the scale, taken "as the crow flies":—

Twenty words (address up to nine words not counted) to
any distance not exceeding 100 miles .... 1s. 0d.
Do. beyond 100, and within 200 miles .... 1s. 6d.
Do. beyond 200 miles.. 2s. 0d.

This scale is very similar to the rates in France, Prussia, &c. The message charge in France is one franc between two places in the same department, or two francs between two places in different departments. The franc rate does not, therefore, on an average, allow a message to be sent more than about fifty miles. As the addresses are counted and charged for on the French system, the English rates, on an average, allow senders an advantage over those of France.

In Switzerland, the internal rate is a franc, but the distances are so small that the average of messages would fall within the 100 mile radius; and as the same system of counting addresses is followed, the charge is practically about the same as in this country.

Belgium for some time adopted the franc rate, but has recently introduced an internal half-franc rate, which doubtless proves a great boon to the inhabitants of that country, though not likely to yield a profit to the State. The distances in Belgium are, however, very small, and the great majority of messages must come within forty or fifty miles. Both Switzerland and Belgium derive great part of their receipts from international (or through) messages passing to and from other countries, and upon which they charge a much higher rate than is applied to the internal traffic. The Swiss returns for 1865 give the following results:—

	No.	Receipts	Average per Message.
Internal messages	364,118	381,378fr.	lfr. 12c.
International	196,377	345,186fr.	lfr. 75c.

The total number of messages was, therefore, 560,495, producing 726,564 francs. The expenses of working and maintaining the

Swiss lines for 1865 amounted to 657,533 francs; so that the receipts from internal messages, apart from the international traffic, would amount to but little more than half the expenses.

The same remarks apply to the Belgian telegraphs, where considerable profits have been derived from the large amounts charged upon the international messages.

In France the telegraph has been worked at a heavy loss up to the present time. The French budget for 1862 showed the loss at that time to be at the rate of 120,000*l.* per annum, but this has been somewhat reduced latterly. As many of the continental telegraph systems are worked in connection with the Post Office departments, it is difficult to arrive at any reliable figures as to the expenses connected with the wires and stations. There is no doubt, however, that the minor stations may be worked in connection with the local post-offices much more economically than when separately established, as in this country.

From comparisons instituted, it is found that the telegraphic communication in the United Kingdom is far more regular, rapid, and correct than those of other countries. The percentage of errors is exceedingly small; and a very large proportion arise from the indistinct writing of the senders, who are frequently excited or hurried in preparing their communications, and hence write carelessly. In many cases the addresses given are insufficient or incorrect, and notwithstanding the companies' efforts, the messages in such cases frequently cannot be delivered. When this happens, the senders are communicated with, if they can be found, and better addresses obtained.

By far the greater proportion of the complaints received from the public relate to messages transmitted over foreign lines, and especially to stations in Turkey, India, Spain, and Portugal; where the local telegraphic arrangements are most imperfect, coupled with an inefficient staff.

In this country, as might be anticipated, the service is worst done at the minor railway stations, where it would not pay the telegraph companies to place a special clerk, and at which the apparatus is worked by the railway company's staff for the convenience of those resident in the locality. Complaints are scarcely ever made in connection with the lines worked by the trained staff of the telegraph companies.

200. The following statistics, showing the progress of the telegraphs in the United Kingdom, may prove interesting:—

Year	Miles of Line	Miles of Wire	No. of Stations	No. of Messages
1850	1,684	6,730	180	29,245
1855	7,649	38,028	586	882,360
1860	10,854	51,556	1,032	1,863,839
1865	16,066	77,440	1,882	4,650,231

During the past year, 1866, the number of stations open in the United Kingdom will have amounted to nearly 2000, and the number of internal messages (exclusive of those to and from foreign countries), transmitted for the public will have exceeded five millions.

It will be seen that the successive reductions of rates and extension of lines have vastly increased the number of messages; but, notwithstanding this, the accounts of the companies show that the lines would not have yielded any fair return upon their outlay if it had not been for the receipts from foreign messages, and contributions in the shape of rents for wires and apparatus from railways, and from the Press for news transmission.

In support of this, The London District Telegraph may be instanced; the wires of which radiate to all the suburbs of the Metropolis, and have been worked at a very low tariff scale of fourpenny and sixpenny messages for the last seven years. Notwithstanding the great facilities this has afforded for communicating between nearly 200 stations in and about London, the company has not been able to make any profits, although the number of messages on its wires have increased from 74,000 in 1860, to 316,000 in 1865.

In contrasting the operations of the Telegraph and the Post Office, it should be borne in mind that each message engages a wire for an appreciable time and requires a separate messenger to deliver it; and also, that each length of telegraph requires a staff to look after, repair, and renew it from time to time.

201. It will be seen from the following table of the number of stations open in foreign countries at the end of 1866 that this country contrasts favourably as regards accommodation to the public:

	Stations		Stations
France	1,550	Prussia	822
Austria	837	Italy	558
Belgium	288	Switzerland	244
Spain	215	Russia	185
Norway	70	Portugal	82
Sweden	121	Turkey	78
Denmark	55	Algeria and Tunis	54

Besides these, other German States, including Baden, Wurttemberg, Bavaria, Saxony, Hanover, Holland, &c., number 1023 stations.

The total number of telegraph offices open on the Continent is, therefore, 6182; and, adding those of this country, the European system comprises slightly over 8000 stations in communication with one another by means of about 85,000 miles of line, with 230,000 miles of wire.

202. In the United States, owing to the great distances between important towns, and the consequent slowness of postal communication, the telegraph made very rapid progress. By 1852 there had been erected 24,375 miles of wire. To 1854, 41,392 miles, and at the end of 1859 about 55,000 miles. During the four years of civil war in the States, the progress of the telegraph was in a great measure suspended; but since then it has again made rapid strides, and at the present time the wire mileage is about the same as in this country, and the mileage of line much greater.

At first the American lines were worked by a number of small independent companies, each owning 400 or 500 miles of telegraph. Many of them did not pay, and it was soon discovered that the mere local traffic was not profitable, and required to be supplemented by through messages. After a few years the lines, which had been in many cases hastily erected, required heavy repairs, which some of the companies could not well afford. These circumstances led to amalgamation in many cases, and a general union of interests between the companies. The rates, which were unremuneratively low on some of the lines, were then adjusted. At present the tariff in the United States is nearly twice as much for a message of ten words as it would cost to send one of twenty words in this country. Thus, between New York and Boston the charge for ten words is 60 cents, or 2s. 6d.; and to Baltimore 70 cents, or 2s. 11d. Both these places are about the same distance as between Liverpool and London, on which the charge for twenty words is 1s. 6d. Again, the charge from New York to Portland is 90 cents, or 3s. 9d., while the rate in this country for twice the number of words transmitted a similar distance is 2s.

203. Stupendous as have been the projects actually realised in this application of science to the social uses of the United States, the extension to California, about 2400 miles in length from the Mississippi across the wildernesses of prairie land, is perhaps the most remarkable. This line, projected and surveyed in 1850, and brought before Congress in 1851, was completed and opened a few years ago. It is carried to the southward, in order to avoid the snow and ice of the northern route over the Sierra Nevada, by the way of the South Pass, crossing the Sierra Nevada Mountains in latitude 39°. The whole distance from the Mississippi to San Francisco is about 2,400 miles.

In a commercial point of view, the line became of great importance, not only as a means of communication between the opposite extremes of a single country, but as the channel for imparting knowledge between distant parts of a vast continent. With the previous postal transmission it required months to convey information from the sunny climes of the East to the less favoured, in point of climate, but not less important regions of the West, teeming as they do with the products of art and enterprise. On this line of wires being established, the Pacific and Atlantic Oceans became as one, and intelligence is now conveyed from London to San Francisco in a shorter time than is required to transmit a letter between Liverpool and London.

Nor does the importance of the undertaking claim less interest when regarded in a social point of view. California is being peopled daily and hourly by our friends, our kindred, and our political brethren. The little bands that a few centuries since landed on the western shores of the Atlantic, have now become a mighty nation. The tide of population has been rolling onward, increasing as it approached the setting sun, until at length our people look abroad upon the Pacific, and have their homes almost within sight of the spice groves of Japan. Although separated from us by thousands of miles of distance, they will be again restored to us in feeling, and still present to our affections, through the help of this noiseless tenant of the wilderness.

This line now links up Vancouver's Island, and by this means our colonists at Victoria were enabled to send their greetings through the Atlantic Cable after it was opened in August last.

Mr. Collins, acting for one of the American telegraph companies, is now busily engaged in endeavouring to carry out a line from British Columbia to Behring's Straits, and thence over Asiatic Russia to the Amoor River in Asia, to which the Russian telegraph has already been extended. If this gap of 6,500 miles is filled up, the belt of wire will then completely encircle the globe.

204. The application of the electric telegraph is not confined to the transmission of messages from one part of the States to another: in the form of a local or municipal telegraph, it is employed as an important instrument of regulation and intelligence in the internal administration of towns.

No adaptation of the system can be more interesting and useful than that which is made for the purpose of conveying signals of alarm and intelligence in the case of fire.

This system was very completely developed in Boston about twelve years ago.

The city is divided into seven districts, each provided with a powerful alarm bell. Every district contains several stations, varying in number according to its size and population. There are altogether in the seven districts forty-two stations. All these stations are connected with a chief central office, to which intelligence of fire is conveyed, and from which the alarm is given; two telegraph wires are employed, a return wire being used to complete the circuit, and provide as completely as possible against accidental interruption or confusion.

At each of the forty-two stations, which are placed at intervals of 100 rods throughout the city, there is erected in some conspicuous position a cast-iron box containing the apparatus for conveying intelligence to the central office. The box is kept locked, but the key is always to be found in the custody of some person in the neighbourhood, whose address is painted on the box door.

On opening this door, access is gained to a handle which is directed, by a notice painted above it, to be turned slowly several times. The handle turns a wheel that carries a certain number of teeth, arranged in two groups, the number of teeth in one representing the district, in the other, the station; those teeth act upon a signal key, closing and breaking the circuit connected with the central office, as many times as there are teeth in the wheel. Signals are thus conveyed to the central office, and, by striking the signal bell a certain number of times, the district and station from which the signal is made is indicated.

An attendant is always on the watch at the central office, and on his attention being called to the signals by the striking of a large call bell, he immediately sets in motion his alarm apparatus, and by depressing his telegraph-key, causes all the alarm bells of the seven districts to toll as many times in quick succession as will indicate the district where the fire has occurred, the alarm being repeated at short intervals for as long a time as may be necessary.

The signal-boxes erected at the stations contain, in addition to the signal-handle, a small electro-magnet, an armature, and a signal-key, so that full and particular communications can be made between each box and the central station, the clicks of the armature forming audible signals.

By this system certain information is given to the central office at the earliest possible moment of the exact locality in which a fire may have broken out, and the alarm is immediately spread over the entire city.

Everyone who is aroused by the alarm, is enabled to tell at once whether interest or duty calls him to the scene of action, and the exact point to which assistance is summoned. Should the alarm be given in the night, those whose attention is awakened may ascertain, from the tolling of the bell, the precise- quarter in which danger threatens, and should they have been needlessly disturbed, may rest in peace, and find, in the knowledge that they and theirs at least are in safety, a consolation for broken slumbers.

The Police Offices and Fire Brigade Stations are now brought into communication with one another, both in London and Liverpool, by somewhat similar arrangements; and Waterworks have also found it desirable to have the telegraph, to give notice directly any leakage shows itself along the main line of pipes, or in reservoirs at a distance.

BELGIAN TELEGRAPH LINES

205. Although in the extent of its territory Belgium is one of the least considerable of the Continental States, it derives from its position in relation to this country, much importance, so far as regards telegraphic communications. By the submarine cable between Dover and Ostend, or failing that, by the cable between Dover and Calais, Belgium constitutes the most direct telegraphic route to the Rhine district and centre of Germany.

The Belgian telegraph lines, as well as the railways, are constructed, maintained, and administered by the State. Separate systems of conducting wires are appropriated to the service of the railways, which is performed exclusively with the alphabetical apparatus of M. Lippens, already described (**155**). There are a few exceptional cases on branch lines of railway, upon which the State has not yet constructed telegraphs for the public service, where private despatches are sent by the railway telegraphs, but generally an extensive system of independent wires, with their accessories, are adapted to this purpose, for which a large corps of telegraphists has been formed.

The gross annual receipts to the end of 1864 had been sufficient, after defraying the working expenses and cost of the lines, to leave a profit of 48,240*l.* upon the 14 years working to that date.

Immediately on the completion of the submarine cable between Dover and Ostend, an active daily intercourse between London and Brussels commenced, and has since been sustained. The connections were completed on the 20th of June, 1853, and on the 27th of the same month 111 despatches were interchanged between the two capitals.

It is proposed to construct wires and apparatus sufficient to maintain the communications on this important line, so that even with the greatest pressure of business, the public shall not have reasonable ground of complaint on account of delay. "A telegraphic line," observes the Minister of Public Works, "should not be organised with the mere powers which suffice for the ordinary or average business, but should be such as to meet the exigencies of occasional pressure, without subjecting the public to delay, or interrupting other regular business. Besides which, it ought never to be forgotten, that in telegraphic business great pressure must always come at particular hours, when prompt expedition is indispensable. This will be easily understood in the business of the Belgian lines, which constitute the route upon which the quotations of the money markets of all the great centres of affairs — London, Paris, Amsterdam, Berlin, Antwerp, &c. — are transmitted at certain hours."

The business transacted by the Belgian telegraphs consists of three classes of despatches:–

HOME DESPATCHES, being those transmitted between two Belgian stations.

INTERNATIONAL DESPATCHES, being those between a Belgian and foreign station.

FOREIGN DESPATCHES, being those transmitted through Belgium in passing between two foreign stations.

Of these three classes of telegraphic business, the second has proved to be the greatest in number, and the third the most productive.

It appears that a very large proportion of the despatches transmitted and received in Belgium, are interchanged with foreign countries, or matter passing *en route* between foreign places. Nearly half the gross amount received for telegraphic despatches has been produced by despatches transmitted between foreign stations, and only passing *en route* through Belgium. This is explained by the fact that such despatches passing always from frontier to frontier, and in the majority of cases from Ostend to the Prussian frontier, the entire length of the kingdom, pay for the longest class of telegraphic distance. This is one of the advantages which the Belgian telegraph derives from the geographical position of the country.

To show the proportion in which the telegraphic service is shared by different subjects of correspondence, we shall take the following classified subjects of despatches:—

	Number	Per cent. of Total
Commerce	3247	56
Money market	1566	27
Private	754	13
Press	116	2
Government	116	2
Total	5799	100

In relation to length the proportion was as follows:—

	Number	Per cent. of Total
From 1 to 20 words	4741	81.8
From 21 to 50 words	921	15.9
From 51 to 100 words	122	2.1
Above 100 words	15	0.2
Total	5799	100.0

Thus it appears that commerce and the Stock Exchange supply 83 per cent, of the whole telegraphic business, 13 per cent. being personal and domestic, and the Press and Government each employing the insignificant proportion of one despatch in every fifty.

It is also apparent that a very small proportion of the despatches exceed the length of twenty words, and almost none that of fifty words.

The large proportion of international and foreign despatches transmitted upon the Belgian wires, and the necessity of pre-payment for despatches, in all cases, to their ultimate destinations, rendered it necessary for the Belgian administration of tele-graphs to make some general arrangement with the principal contiguous states, for such an interchange of correspondence.

The tariff for single messages crossing the Channel, by the Ostend submarine cable, is 4s. to London, or 5s. to any station beyond London.

From 1850 to 1862 the internal tariff was 1½ franc. From 1862 to 1865, 1 franc; and since 1865, half a franc.

At the stations on the Belgian lines the Morse telegraph, as used in the German States, is provided.

The Belgian lines receive and transmit despatches written, at the option of the sender, either in French, German, or English, at all the principal Belgian stations.

Despatches transmitted between Holland and Belgium can be transmitted and received in Dutch, and all despatches between Belgian stations may be sent in Flemish. At all stations despatches are transmitted and received in French.

The number of internal messages has increased from about 80,000, in 1860, to 332,700 in 1865.

FRENCH TELEGRAPHIC LINES

206. The French system is everywhere erected upon posts chemically injected to insure their durability, and there are nowhere less than two conducting wires; but a greater number between all stations where an active correspondence is maintained.

The instruments used for the transmission of despatches are on the Morse system. A few of Hughes's printing instruments have recently been introduced, and are worked on some of the principal circuits from Paris.

The French telegraphic lines communicate with those of England at Calais, Boulogne, and Dieppe by three distinct submarine cables, containing altogether sixteen wires; with those of Belgium at Lille and Douai; with those of Prussia and Northern Germany, at Metz; with the Rhenish States, Wurttemberg, Bavaria, and Austria, at Strasburg; with those of Switzerland, at Mulhouse and Macon, the former communicating with Basle, and the latter with Geneva; with those of Savoy and Piedmont at Grenoble; and, in fine, with those of Spain at Bayonne; and thus the lines are continued to the Spanish frontier, and lines of wire now extend to all the principal places in Spain and Portugal.

In practice the transmission of despatches is not always so direct as it would appear to be upon the inspection of a telegraphic map. Thus, by the three submarine cables, Paris is in permanent direct communication with London. But when it is desired to transmit a despatch from Paris to any of the provincial towns of England, the despatch is at present received and written down at the central station in London, and then repeated and transmitted to the place of its destination in the provinces. This repetition could of course be avoided, by uniting, in the London station, the wire from Paris with the wire leading to the provincial station to which the despatch is addressed, and if the despatch were one of extraordinary length, this course would be the most expeditious; but to adopt it with the ordinary class of short messages, would involve much inconvenience and more delay in general than is incurred in its repetition and retransmission. Thus, to send each message direct to its destination in the provinces, it would be necessary that, previously to the transmission from Paris, notice should be transmitted to London to connect the Paris wires with those between London and the place of destination, and as this change would have to be made separately for every provincial message, and as the wires between London and the various provincial stations must necessarily be occupied, more or less, at all times, in the transmission of home correspondence, the business of transmission in this direct manner would not only be far more dilatory than the process of repetition, but would, in fact, at busy times of the day, be totally impracticable.

What has been here stated respecting the Paris and London line will be applicable, *mutatis mutandis*, not only to all international messages, but in many cases to messages transmitted between home stations, which it is often more convenient and expeditious to repeat and retransmit at certain, intermediate stations than to send direct by the connection of the wires at those stations.

The French Government has carefully organised the administration of the telegraphs throughout its entire territory, and besides modifying and reducing the tariff from time to time, it has placed the whole upon a more efficient footing. It now constitutes an important department of the state, placed under the superintendence of a director-general, four inspectors-general, twelve chief directors, and a hundred inspectors. The director-general, established in Paris, holds his office under the Minister of the Interior, and has authority over all the inferior functionaries. The four inspectors-general control and direct under him the entire telegraphic service throughout the empire. These inspectors, aided by scientific men nominated from time to time by the Minister, form a superior council, charged to consider and decide upon all improvements proposed to be made in the processes, or in the telegraphic apparatus.

The telegraphic lines are distributed into distinct systems or sections, over which the chief directors preside, so as to inspect, direct, and by communication with the inspectors-general and director-general, to centralise the service.

The hundred inspectors will each be charged with the direction of one or more stations, and will have under their authority deputy station-masters, telegraphists, surveyors, artisans, and labourers, charged with the maintenance of the apparatus, the conducting wires, posts, and all the accessories of the line.

In all chief places, the bureaux will be open night and day. The number of stations open on 1st November, 1853, was only 78; and at the end of 1866, no less than 1550.

The posts, a large proportion of which had not sufficient magnitude and strength to bear the necessary number of wires, have been everywhere replaced by others of suitable dimensions, and the telegraphists are augmented in number, and measures taken to ensure their efficiency.

AUSTRO-GERMANIC TELEGRAPH UNION

207. The electric telegraph had not been long in operation in the German States before it became apparent that great inconvenience and much obstruction to the progress of correspondence arose from different states adopting different telegraphic instruments and signals. The difficulties arising from this cause became at length so great as to demand prompt and effectual remedy. A telegraphic congress was accordingly convened at Vienna, in October, 1851, at which deputies from all the German States attended; and after a full discussion of the subject, it was resolved to form an Austro-Germanic Telegraphic Union. This union includes all the states of Europe east and north of the Rhine. It was agreed that a common system of telegraphic instruments and symbols should be adopted throughout all the associated states, and that Morse's telegraph, with its receiving magnets, registers, and uniform alphabet, should be everywhere used, so that telegraphic communication might at all times be made between any two stations of the Union without the delay and inconvenience of translating despatches at intermediate stations from one system of telegraphic symbols into another.

Despatches are transmitted and received at all the stations of the Union, either in German or French. They are also transmitted and received in English at such of the chief stations as are found by experience to have frequent communication with this country.

Notwithstanding the dense population and active commerce of the kingdom of the Netherlands, its limited territory has rendered a very small telegraphic network sufficient for its purposes. They are connected at the Hague by ten submarine wires with the English lines, at Antwerp with those of Belgium, and at Arnhem with those of the German Union.

Despatches are received in German and French at all the stations, and in English at the principal bureaux.

THE SWISS TELEGRAPHS

208. The natural difficulties opposed to the construction of railways in Switzerland did not offer such serious impediments to the construction of telegraphic lines, an extensive network of which has been constructed and brought into operation. Thus Berne is connected with the French lines by wires to Besancon, and with the German lines at Bale. Lausanne is connected with Besancon by an independent line, and also with Berne on one side and Geneva on the other. Geneva is also connected with the French system at Macon, and with that of Savoy at Aix, from whence a line of wires is carried across Mont Cenis to Turin.

From Lausanne the wires are carried by Vevay and Sion through the Valais to the foot of the St. Gothard, across which they are continued by Bellinzona to Milan.

Another line passes from Basle by Lucerne, Glaris, and Coire, to the Splügen, which it crosses, and is carried to meet the former line at Bellinzona, and thence to Milan.

Another line from Basle passes by Zurich and St. Gal to Innsbruck, from whence it passes by Batzen and Trente to Verona, and by Salzburg and Linz to Vienna.

ITALIAN TELEGRAPHIC LINES

209. Italy is put in electric connection with the more northern countries of Europe at six points, Nice, Mont Cenis, the St. Gothard, the Splügen, the Tyrolese Alps, through Innsbruck, and by Trieste.

The French and Swiss lines are connected with Turin by the wires over Mont Cenis already mentioned; the Swiss and Rhenish lines, with Milan, by the wires over the St. Gothard, and the Splügen and the Austrian and Bavarian lines by the wires over the Tyrolese Alps, and those from Trieste round the shores of the Gulf of Venice.

From Venice to Milan a line is carried by Verona and Brescia, which is continued to Turin. From this line there are two branches going southwards, one from Verona by Mantua, Parma, Modena, Lucca, Leghorn, and Florence; thence the wires stretch southwards to Rome, Naples, and the principal southern Italian towns; and are linked up with those of Sicily by means of a short submarine cable across the Straits of Messina. A cable laid in 1859 connects the Sicilian telegraphs with Malta, and thence with Alexandria in Egypt.

OTHER LINES

210. Extensive systems of telegraph are also erected throughout Turkey, Persia, India, and Australia. New Zealand, the Cape of Good Hope, and the various States in South America have also availed themselves of the wires; so that at the present time it may be said that wherever there is civilization, there also may be found the telegraph. In round figures, the world is now overspread by about 160,000 miles of line, comprising 400,000 miles of wire, working between nearly 14,000 stations, and giving employment to a staff of probably 40,000 persons.

If you found this book on the
Electric Telegraph
interesting
you might like
these other titles in
The Electric Telegraph Series

Also available in the
Kindle Store
in e-pub format
as
Kindle Print Book Replicas

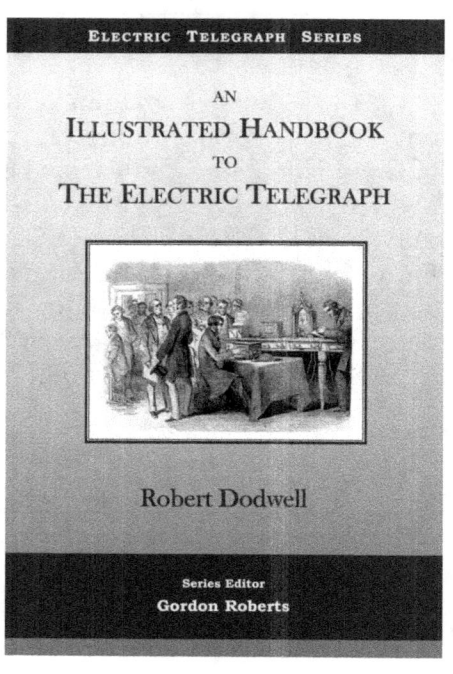

Title:	An Illustrated Handbook to the Electric Telegraph
Author:	Robert Dodwell
Date:	1862
Edition:	Second
Pages:	102
Format:	Paperback
Size:	6 x 9 in (152 x 229 mm)
ISBN:	978-1-9792525-6-0
Series:	Electric Telegraph 1
Pub:	30 Oct 2017

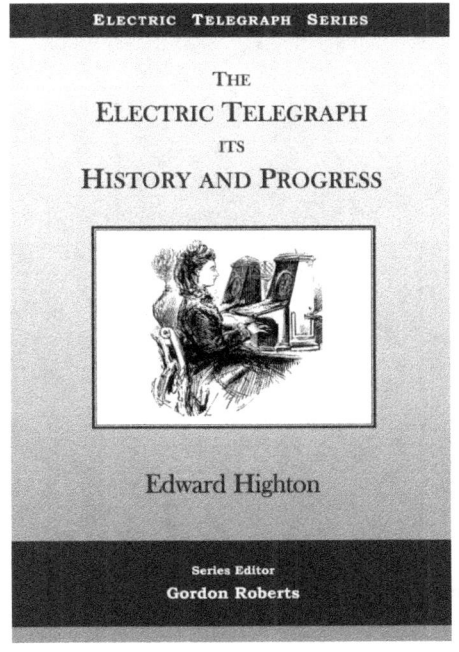

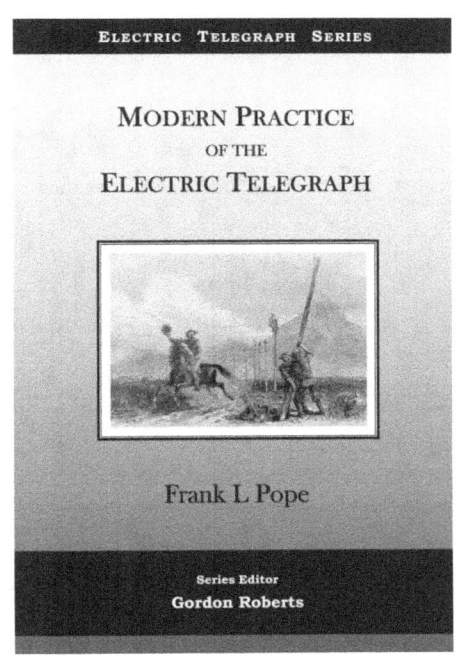

Title:	The Electric Telegraph - Its History and Progress	Title:	Modern Practice of the Electric Telegraph
Author:	Edward Highton	Author:	Frank L. Pope
Date:	1852	Date:	1869
Edition:	First	Edition:	Second
Pages:	200	Pages:	162
Format:	Paperback	Format:	Paperback
Size:	6 x 9 in (152 x 229 mm)	Size:	6 x 9 in (152 x 229 mm)
ISBN:	978-1-9791199-9-3	ISBN:	978-1-9818047-1-9
Series:	Electric Telegraph 2	Series:	Electric Telegraph 3
Pub:	26 Oct 2017	Pub:	18 Dec 2017

RENASCENT BOOKS
*dedicated to creating facsimiles
and reprints of treatises and
texts of great historical
interest.*

www.ingramcontent.com/pod-product-compliance
Lightning Source LLC
Chambersburg PA
CBHW072012230526
45468CB00021B/1196